Animal Geography

By the same author

ELEMENTARY GENETICS (Macmillan)

Animal
Geography

by

WILMA GEORGE, M.A.

Fellow and Tutor in Zoology at Lady Margaret Hall and
Lecturer in Zoology at Somerville College, Oxford

HEINEMANN

LONDON

Heinemann Educational Books Ltd
LONDON EDINBURGH MELBOURNE TORONTO
SINGAPORE JOHANNESBURG AUCKLAND IBADAN
HONG KONG NAIROBI NEW DELHI
KUALA LUMPUR

ISBN 0 435 60345 0

First Published 1962

Reprinted 1963, 1964, 1966, 1969, 1972

Published by Heinemann Educational Books Ltd
48 Charles Street, London W1X 8AH

Phototypeset by Filmset Limited
Printed in Great Britain for the Publishers
by Morrison & Gibb Ltd, London and Edinburgh

To GEORGE CROWTHER

Preface

Animal Geography has been written for all those who have ever wondered why animals are not the same all over the world, why for example there are elephants and rhinoceroses in Africa and in India but not in any part of the New World. It provides an introduction to the study of the distribution of animals round the world, for those who are starting zoology at school and for those whose knowledge has progressed further but who require a brief survey of the subject of zoogeography. I hope it will be of interest to those who travel from continent to continent and who would like to know how the fauna varies over such long distances and how this has come about. I hope it will be read by all those who are interested in animals and maps.

Because the representation of the distribution of animals round the world is a problem of mapping and because maps are visual reproductions of the shapes of the land, I have made animal geography a visual study. Representational animals of the zoogeographical regions of the world have been given pictorial expression, and an attempt has been made at a visual reconstruction not only of the continents during the more recent geological ages, but also of the distribution of a few important terrestrial mammal and other vertebrate families during that time. I believe that this method will give an idea of past land movements and past animal movements in a more easily understood form than can be done by words alone. It involves a certain arbitrariness. Where words can modify a statement to make plain a point of doubt, a drawing, unless it is to be complicated to a state of uselessness, has to make a decision.

There is little agreement amongst paleogeographers over the state of the land in past ages and whatever picture is drawn there will be some who do not approve of it. At best only a rough indication can be given of continental coastlines and land connexions millions of years ago so that although I have based my maps on recent authorities they should not be considered to be either accurate or definitive. Similarly the outer form of the fossil animals as I have drawn them is necessarily imaginary, but their time and place of occurrence is based on up to date evidence and in particular on the fundamental work of G. G. Simpson.

In describing the present day distribution of vertebrates by regions I have enumerated in full only the families of land mammals which occur (excluding the bats), picking out for further description those which seemed to me to be of special interest for my purpose. Anyone who wishes to plot the distribution of those mammalian families which I have not discussed in detail can do so from the information I have provided.

The animals chosen for illustration are not drawn to scale nor are they labelled, but are to be regarded as symbols of the different regions and different epochs. The text makes clear the type of animal that is depicted, but for those who require more precise information on regional animals a list of their scientific names is given at the back of the book.

I would like to thank D. L. Ride for his encouragement and for reading the manuscript at an early stage.

Contents

x

Part One

PRESENT

I

Introduction

Maps

Animal geography, or zoogeography, combines animals with maps.

Maps represent pictorially the surface of the earth. The best guide to the physical characteristics of the world is a sphere, but because of the necessity of representation on a flat page, various methods have been devised for spreading out the sphere so that it can be looked at in two dimensions. The best known method is Mercator's projection which is the flattest looking of all and which is useful for seeing the main features of the world drawn compactly into a rectangle. Because of its obvious utility, it is a projection which is frequently used in spite of its great fault, which is to exaggerate the area of the polar regions. The lines of longitude are drawn as parallels, with the poles at infinite distance, so it is impossible ever to show the poles themselves on a Mercator projection.

Among many other methods, conic projections give some indication of the roundness of the world and the smallness of the polar regions by projecting the lines of longitude on to a cone in such a way that only one parallel touches the surface of the sphere and the apex of the cone is above the surface. The parallels are drawn as circles, the meridians are projected to meet the apex

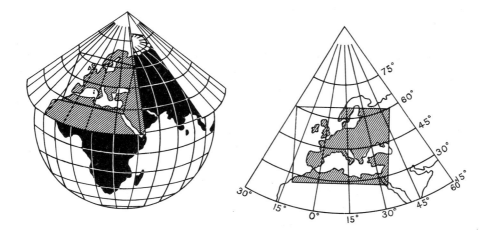

of the imaginary cone. Conic projections are useful when a picture of only a small area of the world is required at any one time. The whole world cannot be represented in continuous view in a conic projection. But a modification of this projection is particularly useful for depicting polar regions. Here the pole becomes the centre of a circle instead of the apex of a cone, the meridians radiate from the centre and the lines of latitude form concentric circles.

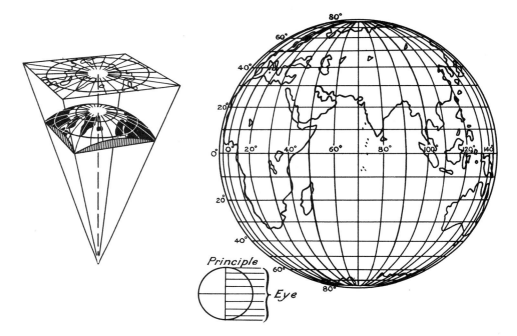

Orthographic projections represent the world in a circle. The plane of projection is the centre of the earth and, in the simplest type, the eye is supposed to be looking at the earth from an infinite distance. This projection suffers like the conic from the impossibility of showing more than half the world in one continuous picture.

Others picture the world within an oval. It has the advantage of bringing the polar regions to a point and of presenting the whole world in one view, but the representation suffers from serious angular distortion of the edges.

All these projections have been used in zoogeography, and have been the cause of considerable confusion and misunderstanding, Mercator's projection being the most commonly used and the most misleading. But as no satisfactory alternative has been found that fits the printed page neatly and shows both the whole world at one time as well as accurately representing the relations of its parts, no one projection has been adhered to in this book. Convenience of fitting the greatest area of land into as small a space as possible has had to be balanced against the possibility of misunderstanding. Convenience and its familiarity have frequently tipped the balance in favour of Mercator's projection, particularly when a continuous picture of the whole

world has been required. For smaller areas and where Mercator is positively
misleading, conic or orthographic projections have been used.

In all cases where the spherical world is represented on a flat page, some
piece of the world is in the middle of the picture and other pieces on the edges.
The choice of the midline is arbitrary, depending often on the country of
origin of the map, though conventionally it is usually either the 0° meridian
or the 180°. The choice of an unsuitable midline may on the one hand confuse
by its unfamiliarity and on the other mislead the zoogeographer. It cannot be
emphasized too strongly, therefore, that in the study of animal geography a
globe is the best guide, and the only working model that cannot be misleading.

Animals

Land covers about 29 per cent of the earth's surface and is unequally distributed. There is more than twice as much land in the northern hemisphere
as in the southern. Most of the land in fact forms a nearly continuous mass,
all the main continents being linked either by continuity of the land or by
archipelagos and islands.

On this land and in its rivers live both vertebrate and invertebrate animals.

Theoretically animal geography is concerned with all animals, the invertebrates and the vertebrates, the terrestrial animals and the aquatic. There are
some 1,000,000 species of animals and therefore any one book is by necessity
selective. Only the geography of land and freshwater vertebrates, about 2 per
cent of all animals, is discussed here and even amongst these there has been
further selection to bring the emphasis on to mammals. For an account of the
seas, the reader is referred to Ekman (1953) and for maps showing the
geography of invertebrates to Bartholomew, Clarke and Grimshaw (1911).
All land and freshwater vertebrates are dealt with in detail in Darlington
(1957).

Animals are grouped into units and this grouping is the basis of classification, the giving of names to different sorts of animals and the indication of
their relationships. Vertebrates are classified as a large group of the phylum
Chordata. The vertebrates themselves are divided into five classes, the fish, the
amphibia, the reptiles, the birds and the mammals. Each class is divided into
orders. The class Amphibia, for example, has three living orders, the Anura
(no tail) frogs and toads, the Urodela (tail visible) newts and salamanders, and
the Apoda (no feet) worm-like caecilians of the tropics.

The orders are further subdivided into grades, one of which is the family.
The family is further divided into sub-families, genera and species. Thus the
giraffe belongs to the class Mammalia, the order Artiodactyla (even toes), the
family Giraffidae, the sub-family Giraffinae, the genus *Giraffa* and the species
camelopardalis. According to the binomial convention the scientific name for
a giraffe is *Giraffa camelopardalis*. Often there is more than one species in a
genus, genus in a family and so on. For instance *Okapia johnstonii*, the okapi,
is in the same family as the giraffe, Giraffidae, but belongs to a different subfamily, Paleotraginae, and genus, *Okapia*. To take another example, in the
horse genus *Equus* (family Equidae), *Equus asinus* is the African species of

wild ass, but *E. caballus* is the domestic horse.

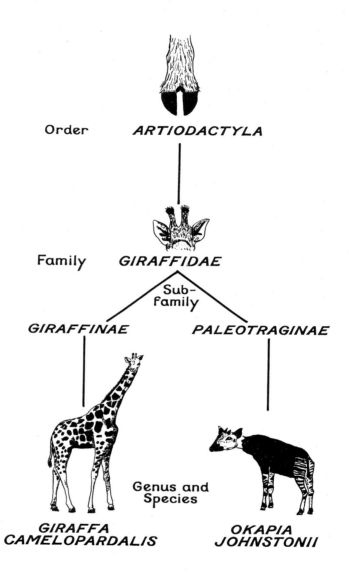

In the account of animal geography that follows the emphasis is on families of animals rather than on orders or species. This is an arbitrary choice but it has been used often and is convenient. The families are the units of classification which best show how vertebrates are spread through the world and there are frequently well-known names to describe them. Thus the word cats is generally recognized to refer to the cat family Felidae, dogs to the family Canidae.

For similar reasons, convenience and limitation, the theories outlined in later chapters make reference mainly to mammals. Mammals are familiar, well classified, well collected, have a well-documented history, and are mainly confined to the land.

The living mammals of the world are classified into families in the following way:

CLASS MAMMALIA

Order Monotremata: egg-laying mammals; two living families.
    F. Tachyglossidae: spiny anteaters
    F. Ornithorhynchidae: duckbilled platypuses

Order Marsupialia: pouched mammals; eight living families.
    F. Didelphidae: opossums
    F. Dasyuridae: 'mice', 'cats' and 'anteaters'
    F. Notoryctidae: pouched 'mole'
    F. Peramelidae: bandicoots
    F. Caenolestidae
    F. Phalangeridae: phalangers, Australian 'opossums'
    F. Phascolomidae: wombats
    F. Macropodidae: kangaroos, wallabies

Order Insectivora: insect-eaters, archaic placental mammals; eight living families.
    F. Solenodontidae: alamiqui
    F. Tenrecidae: tenrecs
    F. Potamogalidae: otter shrew
    F. Chrysochloridae: golden moles
    F. Erinaceidae: hedgehogs
    F. Macroscelidae: elephant shrews
    F. Soricidae: shrews
    F. Talpidae: moles

Order Dermoptera: gliding arboreal mammals; one living family.
    F. Cynocephalidae: colugo

Order Chiroptera: flying mammals; seventeen living families, including fruit bats, vampire bats and the more numerous insectivorous bats.

Order Primates: arboreal omnivorous mammals; eleven living families.

    F. Tupaiidae: tree shrews
    F. Lemuridae: lemurs
    F. Indridae: woolly lemurs
    F. Daubentonidae: aye aye
    F. Lorisidae: lorises, galagos
    F. Tarsiidae: tarsiers
    F. Cebidae: New World monkeys
    F. Callithricidae: marmosets
    F. Cercopithecidae: Old World monkeys
    F. Pongidae: apes
    F. Hominidae: man

Order Edentata: American mammals with simple or no teeth; three living families.

    F. Myrmecophagidae: South American anteaters
    F. Bradypodidae: sloths
    F. Dasypodidae: armadillos

Order Pholidota: scaly anteaters of the Old World; one living family.

    F. Manidae: pangolins

Order Lagomorpha: gnawing mammals with two pairs of upper incisors; two living families.

    F. Ochotonidae: pikas
    F. Leporidae: rabbits, hares

Order Rodentia: gnawing mammals; thirty-two living families, variously grouped into suborders (Wood 1955).

S.O. Sciuromorpha: the squirrel-like forms.

    F. Aplodontidae: sewellel
    F. Sciuridae: squirrels, prairie dogs
    F. Ctenodactylidae: gundis

S.O. Castorimorpha

    F. Castoridae: beavers

S.O. Theridomyomorpha

    F. Anomaluridae: African flying 'squirrels'
    F. Pedetidae: spring haas

S.O. Myomorpha: the mouse-like forms.

    F. Cricetidae: hamsters, voles, rice rats, lemmings
    F. Spalacidae: mole rats
    F. Rhizomyidae: bamboo rats
    F. Muridae: modern rats and mice
    F. Geomyidae: pocket gophers

F. Heteromyidae: pocket mice, kangaroo rats
F. Gliridae: dormice
F. Platacanthomyidae: spiny dormice
F. Seleviniidae: selevinia
F. Zapodidae: jumping mice
F. Dipodidae: jerboas
S.O. Hystricomorpha: Old World porcupines and African rats
    F. Hystricidae: porcupines
    F. Thryonomyidae: cane rats
    F. Petromuridae: rock rats
S.O. Caviomorpha: twelve South American rodent families, including tree porcupines, capybaras, guinea pigs and chinchillas.
S.O. Bathyergomorpha: sand burrowing rodents.
    F. Bathyergidae: blesmols

Order Cetacea: whales; nine living families, all aquatic.

Order Carnivora: flesh-eaters; ten living families.
    F. Canidae: dogs
    F. Ursidae: bears
    F. Procyonidae: raccoons, pandas
    F. Mustelidae: weasels, otters, badgers, skunks etc.
    F. Viverridae: civets
    F. Hyaenidae: hyenas
    F. Felidae: cats
and three families of seals.

Order Tubulidentata: tube-toothed mammal; one living family.
    F. Orycteropidae: aardvark

Order Proboscidea: mammals with trunks; one living family.
    F. Elephantidae: elephants

Order Hyracoidea: one living family.
    F. Procaviidae: hyraxes, coney

Order Sirenia: two living aquatic families.

Order Perissodactyla: odd-toed ungulate mammals; three living families.
    F. Equidae: horses
    F. Tapiridae: tapirs
    F. Rhinocerotidae: rhinoceroses

Order Artiodactyla: even-toed ungulate mammals; nine living families.
    F. Suidae: pigs
    F. Tayassuidae: peccaries
    F. Hippopotamidae: hippos
    F. Camelidae: camels, llamas
    F. Tragulidae: chevrotains
    F. Cervidae: deer
    F. Giraffidae: giraffes, okapi
    F. Antilocapridae: pronghorns
    F. Bovidae: cattle, antelopes, sheep, goats

These are the orders and families of mammals that will be recurring throughout the book and which have, therefore, been set out so that they can be referred to at any time. Certain fossil orders and families will be described as they occur. A complete classification of the mammals can be found in Simpson (1945).

Only comparatively few families of the other vertebrate classes occur constantly and these too are described as they become important. For a comprehensive classification of these other groups, the reader is referred to Romer (1945).

When the geography of animals leaves the broadly geographical level, on the world scale, and turns to what may be called local distribution, in a consideration of islands for instance, the family unit loses some of its usefulness. At this point a few selected genera and species are mentioned. They are described in their place.

2

Animal Maps

Simple Maps

On a map of the world can be marked the area where any family of animal occurs. Such a map shows the range of the family in question, or its geographical distribution, and the map is called a distribution map.

The distribution of family units through the world varies greatly, but when the range of one particular family unit is mapped the simple pattern that emerges falls into one of two main groups. The first illustrates continuous distribution, the second discontinuous distribution. Within both these groups there are gradations from maps of animals with wide ranges to those with restricted ranges.

Of animals that have both a continuous distribution and a wide range there are the rat and bat families amongst the mammals, hawks and cuckoos amongst the birds. These families are world-wide and represent the one extreme of continuous distribution. Less extensive in range though still continuous is the crow family which is absent from New Zealand. Restriction

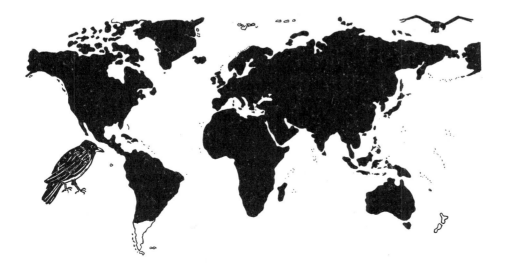

of range may be in an east-west or a north-south direction and many families have a continuous distribution restricted in one of these two ways. Thus the orioles are restricted to the eastern hemisphere, being abundant in the warmer regions of the Old World and Australia but absent from both North and South America. Beavers are restricted in the north-south direction; northern forms, they are absent from South America, Africa, southern Asia and Australia. The simple pattern of carp distribution does not make a clear-cut distinction between these north-south or east-west restrictions. Although it is mainly northern and absent from Australia and South America, it is found in Africa. All these examples illustrate families with a range extending across several continents, but the other extreme of continuous distribution is found in families that are confined to one continent or part of a continent. The giraffe family is confined to Africa and marmoset monkeys to South America. Otter shrews are found only within the equatorial forest belt of Africa. Pronghorns are confined to the north-western prairie land of America.

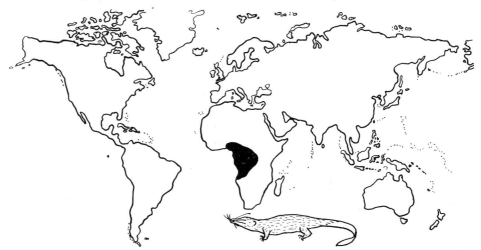

The next map illustrates discontinuous distribution, and again the range may be wide or restricted. The tapir family is discontinuously distributed, occurring in the Malayan area and otherwise only in South America where it ranges widely over the continent. The two discontinuous populations are separated from one another, either by the greater part of the land area of the world, or by the Pacific Ocean, depending on the point that is chosen for the

middle of the map. More restricted in their local range, but also discontinuously distributed over the world, are the liopelmidae a family of primitive frogs. One genus of the family is restricted to a few mountain streams of northwest America and the other occurs locally in New Zealand.

Compound Maps

When the distributions of all animals are added together, the dissimilarities in individual ranges mark out the world into distinct regions. These regions are differentiated from one another by the different mixtures of animals which they contain, as well as by the fact that they may also contain a family which is uniquely found there. For instance one region is characterized by the co-existence of tapirs and members of the camel family even though neither family is itself confined to the region. No other region has this particular combination. In addition this same region has within its confines several families of vertebrates that are found nowhere else in the world.

The world was first divided into these zoogeographical regions on the evidence of its bird fauna. In 1857 Sclater named six avifaunal regions. These regions were only little modified when Wallace (1876) included all the land animals whose ranges were then known, both vertebrate and invertebrate.

Wallace's zoological regions correspond roughly with the continents, each region being separated from its neighbours by some obvious geographical feature, an ocean or a mountain range. The six regions have special names because they do not correspond precisely with political or cultural areas. They are:

> Palearctic
> Nearctic
> Neotropical
> Ethiopian
> Oriental (Sclater's Indian)
> Australian

These regions are generally accepted today because they best combine reality and usefulness.

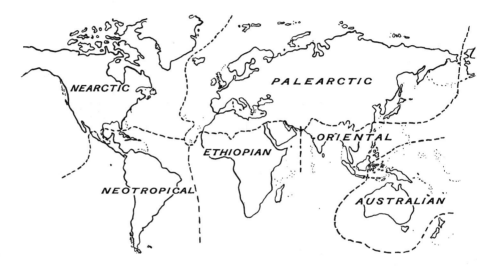

Many other regional classifications have been made, however, some based on the irregularities of distribution of one particular class of animals for which Wallace's regions may not be ideal, others based on the distribution of temperature variations and similar climatic factors. None of these has been found as generally satisfactory as the classification established by Sclater and Wallace in the nineteenth century.

But although generally acceptable, many modifications have been made to the status of the six regions. Some authors have considered the Neotropical and Australian regions to be zoologically so different from the rest of the world and from one another as to rank as regions equivalent to the remaining four put together. In this classification there are three realms, Neogea (Neotropical), Notogea (Australian), and Arctogea (the rest of the world). This grouping of the regions has been found to have some advantages and the names of the three realms are in common use.

Another commonly used modification was suggested by Heilprin in 1887. He proposed that the Palearctic and Nearctic did not merit separate regional status and should be combined into one region, the Holarctic. Many modern authors have accepted this amalgamation of the two northern regions. But, as Wallace himself maintained, it is almost impossible to divide the world into zoological regions all of which have exactly equal status. Some discrepancy can always be found because the regions are the result of natural processes. It is not the purpose of this account to decide on the validity of a region or the suitability of its name, but to outline briefly the zoological characteristics of different parts of the world and to look into the natural processes which have brought them into being. The six regions have therefore been used as the basis and possible differences in status ignored.

In the regional descriptions, the mammal families are listed in full, and these, with a few selected examples of other vertebrates, are used to indicate the general characteristics of each region.

3

Zoogeographical Regions

Palearctic

The Palearctic region is the northerly part of the Old World. It extends over the whole of Europe and the U.S.S.R. to the Pacific coast and south to include the Mediterranean coastal strip of Africa and part of the north of Arabia. It is bounded by the sea to the west, north and east, and by the Sahara and Himalayas to the south. The Palearctic is, therefore, in continuous land connexion with two of its neighbours, the Ethiopian and the Oriental regions, from which it is separated by a desert in the one case and a high mountain range in the other. From its other neighbour, the Nearctic, it is cut off by sea.

The climate is on the whole temperate. The region includes both wet forest lands and dry open steppe land as well as large areas of coniferous forest and a fringe of tundra. The fauna in these various climatic and vegetational sub-divisions of the region also varies. All six regions vary within themselves according to the physical and vegetational differences encountered within their boundaries, but in considering only broad geographical classifications these differences are ignored.

Within the Palearctic there are representatives of twenty-eight families of land mammals, excluding the numerous bat families. This is not as large a number as some of the other regions have, nor does it include any very spectacular animals. But amongst them are families which have a wide con-

tinuous range, a restricted continuous range, a discontinuous range, and two families that are unique. Of Palearctic mammals, rabbit, murid mouse and dog families as well as several bat families are world-wide in their distribution; there are shrews, squirrels, cricetid mice, mustelids and members of the cat family in every other region except the Australian. Other Palearctic families are more restricted and occur in only four of the world's regions. The Palearctic shares bears and deer with Nearctic, Neotropical and Oriental regions, and bovids with the Nearctic, Ethiopian and Oriental. Restricted to the Old World are the families of hedgehogs, hystricid porcupines, civets, hyenas and pigs which occur in the Palearctic, Ethiopian and Oriental regions. In contrast the procyonids (pandas and raccoons) are otherwise known only from the New World. Dormice, jerboas (Dipodidae), coneys and wild horses occur in the Palearctic and in the Ethiopian region, whilst moles, pikas, beavers and jumping mice (Zapodidae) have a northerly distribution and are confined to Palearctic and Nearctic regions.

From this it can be seen that nearly a third of the twenty-eight Palearctic families have a wide range, whilst at the other extreme four families are shared by the Palearctic and Nearctic alone, and another four exclusively with the Ethiopian.

The camel family is the only mammal family occurring in the Palearctic which has a discontinuous distribution. The relatives of the camel, vicunas and guanacos, live in the Neotropical region. It is doubtful whether Palearctic camels are any longer genuinely wild, although some of the two-humped animals from central Asia may be so.

The two families of mammals that are restricted to the Palearctic are both myomorph rodents. They are the Spalacidae or mole rats and the Seleviniidae. Each family is represented by only one genus. *Spalax* is a brownish-yellow burrowing animal with no tail but typical rodent gnawing incisors. *Selevinia* is mainly remarkable for not having been discovered until 1938. Even then the first specimens were of skeletons left by vultures in Kazakhstan, and it was only after this discovery that the living animals were found.

Almost all the birds of the Palearctic belong to families which are of very wide distribution. There are pheasants, wrens, blackbirds, finches, warblers, sea birds, geese and birds of prey amongst the very large total number of families. There are no parrots. The only family that is restricted to the region is the hedge-sparrow family. The Palearctic shares its bird families, like its mammal families, with one or more of the neighbouring regions, in the main either with the Nearctic or with the Old World tropical regions, the Ethiopian and the Oriental.

As for the reptiles, the Palearctic region is more conspicuous for the absence from it of those which are found in other regions, than for the presence of distinctive forms. There are a few lizards, snakes and tortoises, and an alligator in China. No family is confined to the region.

In contrast, the Palearctic has a large number of tailed amphibia, ranging from the common newt, through the black and yellow European salamander,

and the blind white *Proteus* of the Adriatic caves, to the giant salamander from eastern Asia which is $5\frac{1}{2}$ feet long. Indeed most of the tailed amphibia (Urodela) are found either in the Palearctic or the Nearctic regions, across which the families are spread. Of the tailless amphibia (Anura) the common frogs and toads are widespread and there are a few tree frogs of both the hylid and polypedatid (rhacophorid) families.

Carp, salmon, pike, perch and sticklebacks are all common in Palearctic fresh waters. The carp family is the dominant family although it is not restricted to the region but occurs also in the Nearctic and Ethiopian regions.

The vertebrate fauna of the Palearctic is not then very rich. The characteristics of its fauna can be summarized as a complex of Old World tropical families and New World temperate. There are few endemic families.

Nearctic

The Nearctic region covers the whole of North America and extends south as far as the middle of Mexico. It includes Greenland in the east and the Aleutian islands in the west. Except for the narrow strip of Central America it is cut off from all the other regions by sea, although the sea that separates it from the eastern Palearctic is very narrow.

The range of climate and vegetation in this region resembles that of the Palearctic.

Like the Palearctic, comparatively few families are represented in the Nearctic, and of these the largest proportion is made up of those with a wide range such as shrews, rabbits, squirrels, cricetid mice, dogs, mustelids, cats and bats. Bears, procyonids, deer and bovids also have a considerable, though not identical, range. Of the rest of the Nearctic mammals, four families are found otherwise only in the Palearctic. These are the families with a northerly range, moles, pikas, beavers and zapodidae jumping mice. With its other neighbour, the Neotropical, another four families are shared, families which have a westerly or New World range. In each case, only one genus of the shared family occurs in the Nearctic. There is an opossum, an armadillo, a tree porcupine and a peccary. The opossum and armadillo are considered to be comparatively recent immigrants from the Neotropical region, together with the tree porcupine, whereas the peccaries are primarily of northern origin.

Of the twenty-four families of Nearctic land mammals none has a discontinuous distribution. In other words the Nearctic does not share a family exclusively with either of the Old World tropical regions or with Australia.

The fauna of a region is characterized not only by the animals that it contains but also by the absence from it of otherwise widespread or neighbouring families. In the case of the Nearctic, many Palearctic and Neotropical families are not represented. There are, for instance, no hedgehogs, hyenas or pigs from the Palearctic, no tapirs from the Neotropical and,

rather surprisingly, no camels which are represented in both the neighbouring regions.

But although the total number of mammal families is less than that of the
Palearctic and so many families are unexpectedly absent, the Nearctic has
four endemic families. Like the two endemic families of the Palearctic, three
of these families are rodents, the other an artiodactyl. All four live primarily
in the west of the region, the pocket gophers and pocket mice living mainly
in the arid areas of the west and south of North America, the sewellel near
water in mountainous areas of the Pacific coast, and pronghorns on the
prairies of western and central North America. The pronghorn family is an
interesting group of artiodactyls related to both the deer and the bovids,
but differing from both of them in the construction of the short, branched
horns. The horns have a boney core and, like the deer, a soft skin covering,

but unlike the deer only the covering of the horns is shed each year. Sewellels are superficially like muskrats but seem to be related to squirrels, whereas gophers and pocket mice form closely allied families of the myomorph rodents. These two live mainly in the Nearctic and are counted as endemic Nearctic families, but they have in fact spread southwards into the intermediate region where the Nearctic and Neotropical faunas overlap. Intermixture of faunas in this way is common between regions that have land connexions with one another.

To add to the native families of the Nearctic, murid mice and horses, along with domesticated pigs, have been introduced into the region by man.

Nearctic birds are even less differentiated from the neighbouring regions than the mammals. Wood-warblers extend over the whole of temperate North America, and there are several distinctive genera of grouse. Bright red cardinals, tanagers and humming birds are abundant, but they also live in the Neotropical region which is probably their original home. Only the wild turkeys can really be considered to belong exclusively to the Nearctic, and even they have a few stragglers into the Central American parts of the Neotropical.

In contrast to the Palearctic, the Nearctic is the home of many reptiles. Most of the snapping turtles and musk turtles are found there, as well as terrapins. Harmless garter snakes and poisonous rattlesnakes are typical of the Nearctic though not confined to it. There are geckos, horned iguanid lizards and skinks. The gila monster, the only poisonous lizard, belongs to a family exclusive to the region.

Tailed amphibia are abundant, some of them shared with the Palearctic, others more or less confined to the region. Among them are American salamanders, ambystomid axolotls and several of the neotenous forms of urodeles with reduced limbs and, like *Siren*, gills retained into adult life. In mountain streams of the north-west U.S.A. lives the frog *Ascaphus*. Like other frogs and toads it has no tail, but it has tail-wagging muscles, suggesting descent from some tailed ancestor. It shares this peculiarity with only one other frog, *Liopelma*, of New Zealand. The two are therefore considered to be closely related and are included in the same family, liopelmidae. In addition there are common frogs and toads and hylid tree frogs.

A few of the freshwater fish also belong to ancient and isolated families. The garpike and the bowfin are the only living members of a group of fish, the Holostei, which were much commoner millions of years ago than they are today. The garpikes of this group still have the covering of thick scales which was characteristic of all early fish, but which has been lost in modern forms. In addition, the mooneye and bass families are confined to the region and there are many carp and perch.

Compared with the Palearctic, the Nearctic is richer in reptiles and has more endemic families, but in other respects shows considerable resemblance

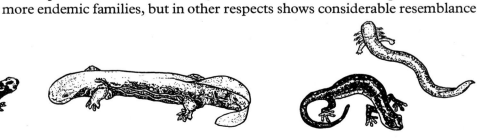

to it. Like the Palearctic its fauna is a complex of tropical and temperate families, but in contrast to the Palearctic, its fauna can be described as a combination of New World tropical and Old World temperate. Since both Palearctic and Nearctic are north temperate and have land connexions with tropical regions, and are the only regions which have these characteristics, their similarities are hardly surprising.

Neotropical

The whole of South America, most of Mexico and the West Indies make up the Neotropical region. It is joined to the Nearctic by the Central American isthmus and separated from all other regions by sea.

The region is mostly tropical, only the southernmost part extending into the south temperate zone. From the west to the east runs the river Amazon with its hundreds of square miles of evergreen forests. Further south the rain forests give place to extensive grassy plains and small semi-desert areas, whilst in the west the long range of the Andes has high mountain forests, plateau land and gentle slopes.

The fauna of the Neotropical region is both distinctive and varied. Excluding bats, thirty-two families of mammals are represented in the region, of which seven are of wide distribution and sixteen are unique. This is the highest number of endemic families for any region.

Like the two preceding regions, the Neotropical has shrews, rabbits, squirrels, cricetid mice, dogs, bears (but only one genus), procyonids, mustelids, cats and deer. It shares with its northerly neighbour, the Nearctic, the opossum, armadillo, tree porcupine and peccary families.

Two Neotropical families show discontinuous distribution, the camel family and the tapirs. The camelids which are otherwise only Palearctic are represented in the Neotropical by one endemic genus, *Lama*, of which there are two species guanaco and vicuna, which live wild on the plateaux of the Andes and on the open grasslands. Two domesticated varieties, llamas and alpacas, have been derived from the guanaco. Differently, but still discontinuously distributed is the tapir family. Although separated by such a great distance, the Neotropical population is considered to belong to the same genus as the Oriental. Three species of this genus, heavily built animals with short tusks and small trunks, live in the forests of the Andes. Tapirs are related to rhinoceroses and more distantly to horses.

Many of the commonly occurring mammalian families have no representatives in the Neotropical region. There are no hedgehogs, moles, beavers, hyenas, native bovids or native horses. The absence of horses is curious for only a few thousand years ago they existed both here and in North America.

But in spite of the absence of common families, the Neotropical makes up its fauna by an abundance of exclusive families. Three families together form

an order of mammals, the Edentata, which is almost confined to the region. Only two other regions, the Ethiopian and Oriental have the distinction of containing unique mammal orders. In the latter cases the orders are small and the animals neither important nor numerous members of the fauna, but in the Neotropical the Edentata are rich in number and variety. Three families, anteaters, sloths and armadillos make up the order, and of these only one genus *Dasypus* the nine-banded armadillo has left the region to colonize successfully the Nearctic. There are three distinct types of anteater, ranging from the large bushy-tailed ground living animal to the tiny prehensile-tailed anteater of the forest trees. They are the only family of the Edentata which conform to the definition implied in the name: without teeth. There are two genera of sloths, both committed to an upside-down life in the trees, and a great variety of armadillos all of which have an armour of small bones covered with a layer of horny material. In spite of this covering armadillos are active burrowing and running animals.

Of the rest of the Neotropical endemic families there are marsupials, two families of monkeys, eleven of caviomorph rodents and, in addition, five families of bats.

Two families of marsupials occur in the Neotropical, but only one of them, the mouse-like Caenolestidae, is confined to the region. Didelphidae, or opossums, are found also in the Nearctic. Apart from these two families, the only other living marsupials occur in the Australian region where there are a further six families.

South American monkeys are diverse; douroucoulis, spider monkeys, capuchins, squirrel monkeys and howlers in the cebid family, and tamarins and marmosets in the callithricid family. The two families have in common the flat-nose feature, which separates them both from the downward-nose Old World monkeys. In other ways they differ greatly from each other. The small bushy-tailed marmosets and tamarins have clawed fingers and toes in contrast to the cebids, more typically monkey with flat nails and prehensile tails.

Eleven specialized families of rodents make the Neotropical the home of the caviomorph suborder of rodents. The New World porcupines are the only caviomorphs that are not confined to the region, occurring elsewhere only in the Nearctic. The porcupines are mainly arboreal and some of them have prehensile tails. Prehensile tails are common in Neotropical trees. The endemic caviomorph families include the largest of all rodents, the capybara which is four feet long, the guinea pigs, spotted pacaranas, pacas, agoutis, chinchillas, nutrias, burrowing degus and tucotucos, the rat chinchilla and the spiny rats.

Finally there are five indigenous families of bats of which the vampire bats are the best known. They are a serious pest of cattle herds, carrying diseases such as rabies, and it has been suggested that it was they who caused the extinction of the native horses of America.

As if this mammalian fauna were not striking enough, the birds equal it in their diversity and strangeness, so much so that South America has been called the Bird Continent. Nearly half the bird fauna is composed of restricted families, and two orders are confined to the region, the rheas or American ostriches and the partridge-like tinamus. Amongst the restricted families are crested toucans with enormous coloured beaks, trumpeters and hoatzins. Cracids to which the curl-crested curassows belong are almost restricted to the region and so too cotingids which include the flaming cock-of-the-rock and the black umbrella bird. On the forested slopes of the Andes

hover brilliantly coloured humming-birds, and in the Amazon forests live macaws, large variegated parrots with long tails. The only conspicuous lack in the bird fauna is a scarcity of song birds, and except for quails, there are no members of the pheasant family.

Snakes, lizards, crocodiles and turtles abound, many of them shared with the Nearctic but others, primarily tropical, with Africa and the Oriental region. Snakes include both constrictors and biters; anacondas, boa constrictors, pit vipers and coral snakes. New World iguanid lizards are numerous and varied, and the small tegus are almost confined to the region. Alligators and caimans are found in South American rivers, but only the caimans are restricted to the region. The Neotropical shares a family of mud-turtles (Pelomedusidae) with Africa, and snake-neck turtles (Chelyidae) with Australia. Both these families are pleurodires, or side-necks, so-called because of the way they bend their necks into the shell. They form a separate suborder from the commoner cryptodire turtles.

Amphibians of the Neotropical are almost all anurans in marked contrast to the two northern regions. There are hylid tree frogs as well as the commoner frogs and toads. The pipid family which is here represented by *Pipa* the Surinam toad, and *Protopipa*, is absent from all other regions except the Ethiopian and its distribution resembles that of the pelomedusid side-neck turtles. *Rhynophrinus*, the sole representative of its family, though related to the pipids, occurs in Central America. It is a pink and brown burrowing toad that feeds on termites. Only one tailed amphibian, *Oedipus*, is widespread through the region but a few axolotls have spread from the Nearctic as far as Mexico.

There are no carp in the region and the fish fauna is dominated instead by the characin fish, gymnotids and cat-fishes. Gymnotids, the electric eels of the Amazon are confined to the region but characins which include the vicious carnivorous piranhas, are, like the pipid toads and pelomedusid side-neck turtles, found also in the Ethiopian region. Less abundant, but equally curiously distributed, is the lungfish, of the Amazon, *Lepidosiren*, related in the same family to an African form. A third occurs in Australia and these three genera are the only members of the lungfish order, the Dipnoi. They have in common, lungs and internal nostrils additional to the gills of other fish, two pairs of lobed fins and a distribution which restricts them to the southern hemisphere.

In sum, the Neotropical is rich in endemic families of vertebrates of all classes, and, of the more widely distributed families, it shares many with the Nearctic and several with other tropical regions of the world.

Ethiopian

The Ethiopian region covers the continent of Africa south of the Atlas mountains and the Sahara, and includes the southern corner of Arabia. Like the Neotropical it has land continuity with its northern neighbour but is otherwise isolated by sea. Also like the Neotropical, it has big rivers and tropical evergreen forests, as well as mountains and grassy plains, but it does not reach as far into the southern temperate zone. The large island of Madagascar with its smaller neighbours is often included in this region but for reasons that will be discussed later it is here treated separately.

The Ethiopian mammal fauna is the most varied of all the regions consisting of thirty-eight families, excluding bats. In number of unique families it ranks second only to the Neotropical.

Only the shrews, rabbits, squirrels, cricetid mice, murid mice, dogs, mustelids, cats and bovids have a wide distribution. Apart from the twelve exclusive families the rest of the mammals are shared with either the Palearctic or the Oriental regions, or, as in the case of the hedgehogs, porcupines, civets, hyenas and pigs, with both.

Thus the Ethiopian shares with the more northerly Palearctic, families of dormice, jerboas (Dipodidae), coneys and wild horses. But it also differs markedly from this region in being without moles, beavers, bears and camels.

With the Oriental region the Ethiopian shares eight of its mammal families, three of which are primates and two, large ungulates. Lorises, Old

World monkeys, apes, pangolins, bamboo rats, elephants, rhinoceroses and chevrotains are all confined to the Old World tropical regions. The pangolins or scaly anteaters looking like large fircones, belong to only one genus, shared by both regions, but all the other shared families are different at the generic level in the two regions. Thus *Loxodonta*, the African elephant, with its huge ears and long tusks is differentiated from *Elephas* the Indian elephant which is an altogether smaller animal. White and black rhinoceroses, each with two nasal horns, represent the Ethiopian genera in contrast to two Oriental genera, only one of which is two-horned. Of the primate lorisidae, pottos and galagos live in Africa, whilst slender lorises are inhabitants of the Oriental region. The abundant Old World monkeys of the Ethiopian region are diverse, and more varied than their relatives of the Oriental. In Africa there are macaques, drills, baboons, mangabeys, guenons and geladas in contrast to a smaller number of Oriental langurs. In the forests of western and central Africa live two of the four great apes of the world, the gorilla and the chimpanzee. The other two orang utan and gibbon are Oriental.

The Ethiopian region has no mammal families exclusively in common with either the Nearctic or the Neotropical.

The African scene is pictured with herds of large herbivorous animals on open plains, zebras, loping giraffes, leaping and springing antelopes, rhinoceroses, elephants and, hidden, waiting for an endless supply of food, lions and other members of the cat family. In fact, of all these African mammals only the giraffe family is confined to the region, although the antelope subfamily of the bovids has reached a diversity in Africa which is found nowhere else.

Including the giraffes, there are twelve unique families of Ethiopian mammals: hippopotamuses and aardvarks; three families of insectivores; and six families of rodents, making an interesting comparison with the Neotropical whose rodents are also diverse and restricted. Like the giraffe family the amphibious hippo family contains only two genera, and the aardvarks, forming on their own the order Tubulidentata, only one genus. The aardvark, or cape anteater, is the size of a small pig with a highly curved back, long snout and long tongue. On the four digits of its front feet and the five of its hind there are sharp hoofs for digging through termites' nests.

Endemic insectivores are otter shrews, golden moles and elephant shrews, and the remaining six endemic families are rodents, whose relationship with the rodents of other parts of the world is obscure. The Anomaluridae and the Pedetidae may be allied to the sciuromorphs although there is no unanimity of opinion on this. Some of the anomalurids are squirrel-like but others are more like mice, some are gliders like the gliding-mice *Idiurus* and the African flying squirrel *Anomalurus*, but some are not. Probably allied to them is the Spring Haas, the sole representative of its family Pedetidae. Of the remaining rodents the cane rats and the rock rats probably have affinities with the Old World porcupines, whereas the gundis and blesmols seem to stand on their own.

Birds are numerous in the Ethiopian region having, like the mammals, strong affinities with the Oriental region and only six exclusive families. There are cuckoos, woodpeckers and hornbills as well as sunbirds, orioles and many birds of prey, but only comparatively few pigeons, parrots and pheasants. Exclusive to the region are ostriches, secretary birds, hammerheads, crested touracos, mousebirds and helmet shrikes. The ostrich is the only member of a unique order, doubtfully related to other large flightless birds in other parts of the southern hemisphere.

Many snakes, including constricting pythons and biting poisonous vipers occur in the region, and amongst the lizards the Cordylidae, or spiny lizards, are restricted, and the chameleon family nearly so. Only four of the fifty species of chameleon are found outside Africa and only one of these lives as far away as India. There are a few agamid and lacertid lizards but no iguanids. Crocodiles and turtles abound, amongst them the Neotropical pelomedusid family of the side-necks.

Amphibians are less distinctive but amongst them, in addition to widespread frogs and toads, is the pipid family represented by *Xenopus*, the clawed toad, which like its Neotropical relatives is aquatic. Hylid tree frogs are absent, their place being taken by another family of tree frogs, the polypedatids. Like other southern regions, the Ethiopian has no tailed amphibians.

The fish fauna, in contrast, is diverse and includes carp, Old World catfishes, characins, lungfish and several endemic families, amongst them the mormyrids. From electric organs in their tails the mormyrids generate an electric field and they are made aware of prey in their muddy pools when this field is distorted. Mormyrids are not related to the electric eels of the Amazon, but the characin fish are a family shared by Ethiopian and Neotropical. Similarly, the lungfish *Protopterus* of Africa is related to the Neotropical lungfish *Lepidosiren*.

The vertebrate fauna of the Ethiopian region is the most varied of all the regions, and in number of endemic families is second only to the Neotropical. In its fish, amphibia and reptiles it resembles both the Neotropical and the Oriental in several ways but in its birds and mammals has overwhelming affinities with the Oriental. Thus the Ethiopian has certain similarities with both Neotropical and Oriental regions because all three have a tropical climate, but its similarities are much stronger with the tropics of the Old World than with those of the New.

Oriental

The Oriental takes in India, Indochina, south China and Malaya as well as the westerly isles of the Malay Archipelago. It is bounded by the Himalayas in the north and the Indian and Pacific oceans on its other sides, but there is no definite physical boundary in the south-east corner where the islands of the Malay Archipelago string out until they reach Australia. The big islands, Sumatra, Java and Borneo, with the Philippine group, certainly belong to the Oriental region, but the other islands are more difficult to place.

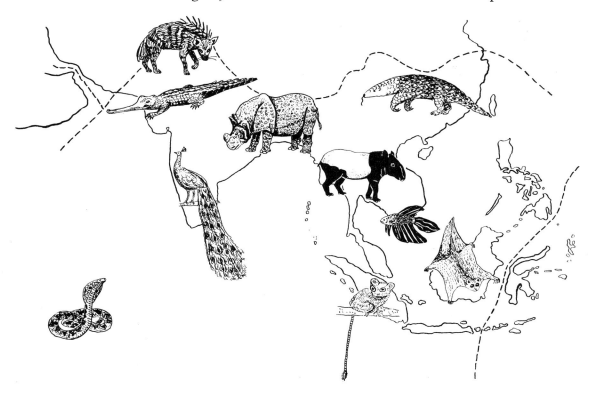

The climate of the Oriental region is mainly tropical.

Although the fauna resembles that of the Ethiopian region it is neither so rich in endemic families nor does it exhibit the same variety of widespread families. Of its thirty mammalian families, excluding bats, only four are endemic, like the Nearctic. One of the four belongs to an endemic order and in this the Oriental resembles both the Neotropical and the Ethiopian.

There are shrews, rabbits, squirrels, cricetids, murid mice, dogs, mustelids, cats and bovids. It shares hedgehogs, porcupines, civets, hyenas and pigs with both Palearctic and Ethiopian regions, but a quarter of its mammal families it shares exclusively with Africa. These are lorises, Old World monkeys, apes, pangolins, bamboo rats (Rhizomyidae), elephants, rhinoceroses and chevrotains. In spite of this considerable resemblance to the Ethiopian fauna, there are striking differences between the two regions.

Unlike the Ethiopian, the Oriental region has moles, bears, tapirs and deer in its fauna. In contrast, it is without jerboas (Dipodidae), coneys and horses. Furthermore it has its four unique families.

The family Cynocephalidae forms by itself the order Dermoptera. There is only one genus *Cynocephalus* (= *Galeopithecus*) the colugo. The affinities of the order are difficult to determine but it is likely to have been derived from the insectivores, diverging from them early in their history. The colugo is a gliding mammal, being aided in its leaps through the trees by a softly furred membrane which stretches from its neck to the tip of its tail and includes both the fore and hind limbs. Only the fingers and toes are excluded from the membrane. At rest on a tree the colugo spreads its membrane and thus, by elimination of its shadow, becomes nearly invisible.

Two primate families are confined to the Oriental region. They are tree shrews and tarsiers, of particular interest in evolutionary studies because they represent crucial stages in the primate story. The tree shrews lie near the base of the primate stock, having links with their insectivore ancestors through the African elephant shrews. Tarsiers represent a more advanced stage of primate evolution belonging to a group which was ancestral to the monkeys. Both these primate families contain small purely arboreal animals, living on an omnivorous diet and using their hands to bring food to their mouths. With these two families and representatives of lorises, Old World monkeys and apes, the Oriental region has half of all primate families, the largest proportion of this order anywhere.

Like the primate families and the colugo, the other unique Oriental family, the spiny dormouse, is also arboreal, but a rodent. Belonging to the myomorph suborder, the spiny dormice live in tall trees in India, boring their own holes in the wood.

Many brilliantly coloured birds live in the Oriental region. Most of them are common to other regions like the widespread woodpeckers or the tropical barbets, but many like sunbirds and hornbills show affinities with the Ethiopian region. Like Africa, the Oriental region has few parrots, but in contrast there are many pigeons and an abundance of pheasants. Indeed the Oriental with the Palearctic is the main home of pheasants. It is the native home of the peacock, the magnificent Argus pheasant and the jungle fowl, from which modern domestic poultry have been bred. With the Australian the Oriental region shares four bird families and has only one exclusive family, the fairy bluebirds.

Lizards, snakes and turtles are plentiful. There is a particularly large number of biting poisonous snakes, vipers and pit vipers, as well as constricting pythons. Lizards include cosmopolitan geckos and skinks as well as Old World agamids and varanids. Crocodiles are widespread and in addition there are gavials, slender-nosed crocodiles, an exclusive fish-eating family confined to India and Malaya. A freshwater turtle, the bigheaded turtle, is also confined to the region and there are many land turtles.

Very few tailed amphibia reach the northernmost parts of the region only, from the Palearctic, but anurans are common. Again like the Ethiopian region the tree frogs of the Oriental belong to the polypedatid family, hylids

being absent from all but a small area of northern Indochina. Pipid toads, however, do not occur, but common frogs and toads range widely.

The Oriental fish fauna is dominated by carp and catfishes.

The many similarities between the Oriental fauna and the Ethiopian arise largely because they are both tropical regions and both situated in the Old World. Although similarities can be found in all groups of vertebrates, they are more pronounced among the mammals and birds.

Australian

The Australian region covers Australia, Tasmania and New Guinea and a few of the smaller islands of the Malay Archipelago, but New Zealand and the islands of the Pacific are not included. The region is unique in having no land connexions with any other region.

The northern part of the region, north Australia and New Guinea, lies within the tropics with high summer temperatures and much of the area is covered by rain forest. The interior of the Australian continent is also hot, but dry, while further south the climate becomes mainly temperate.

The most striking characteristic of the vertebrate fauna of the Australian region is its paucity. But what it lacks in variety and number of families, it makes up for in the uniqueness of many. Apart from bats, there are only nine families of mammals, and eight of these are unique. In addition to these there are rabbits, foxes, rats and mice which are recent introductions into Australia from the Palearctic, and dingo dogs and pigs which are probably also human introductions but dating from prehistoric times.

Of the nine feral families only one is a placental mammal and this, the murid mouse family, is of wide range in the Old World, but Australian murids are of distinct genera and the Hydromyinae water rats an exclusive subfamily.

The dominant mammal fauna is marsupial, made up of six families none of which occurs in the New World, where are found the only other living marsupials.

In a continent which has few placentals, none of them carnivores, the marsupials have become diverse and have taken to ways of life which in other parts of the world are followed by placentals. There is thus a striking parallelism in superficial structure between some Australian mammals and their counterparts in other regions. The marsupial mole, from the family Notoryctidae, resembles placental moles with its paddle-shaped feet and strongly clawed fingers.

The Tasmanian wolf, now probably extinct, looked like a hyena and lived a similar predatory life. In the same family, Dasyuridae, there are pouched 'mice', pouched 'jerboas', 'cats' and 'anteaters'.

The Peramelidae, bandicoots, can be considered the rabbits and insectivores of the marsupial world, whilst the Phascolomidae, wombats, parallel the large rodents.

In the phalanger family there are Australian 'opossums', 'squirrels' and flying phalangers equivalent in all superficial respects to flying placental squirrels.

Finally there are the kangaroos and wallabies which have no exact structural parallels. But although their looks are distinctive, their herbivorous diet and speed of travel in open country, suggest parallels with the ungulates of other parts of the world.

This parallel radiation of the marsupials is not, however, perfect. There are no marsupial bats, seals or whales. Placental representatives of these orders occupy the air and seas of Australia.

The remaining two families of Australian mammals belong to a separate sub-class of mammals, Monotremata. They are the egg-laying mammals and

their relationships with the marsupials and placentals is very remote. They may even have had a separate origin from reptiles, after these had acquired hair but had not yet lost their egg-laying habits. The two Australian families are the only living Monotremes. Duckbilled platypuses form one family, echidnas or spiny anteaters the other. They probably owe their survival as much to their specialized ways of life as to the absence of placental carnivores from Australia. The platypus is semi-aquatic and echidnas are active insect eaters. Both families lay small leathery eggs which the mother incubates, the platypus by curling round them, the echidnas in a pouch. In both families the young are fed on milk which seeps on to the fur from special pores on the underside of the mother.

The bird fauna of Australia does not equal the mammals in peculiarity, for the vast majority of the birds belong to families with a wide range. Trogons and kingfishers, hawks and cuckoos all occur in the region together with pigeons and parrots both of which reach their greatest diversity there. So diverse are the Australian parrots that they are recognized as three exclusive subfamilies, cockatoos, lories and pygmy parrots. Four bird families are shared with the Oriental region; frogmouths, woodswallows, flowerpeckers and megapodes; but there are no pheasants which are so abundant in the Oriental, no finches, no barbets and no woodpeckers. Ten families are unique and include two of flightless birds (cassowaries and emus), honeysuckers, lyrebirds, bowerbirds and the legendary birds of paradise, known to the Portuguese as birds of the sun and thought by them to have no feet.

Australian reptiles are only moderately varied and only two families are exclusive. Constricting pythons and biting tiger snakes are abundant and there are geckos and skinks, agamid lizards and the varanid komodo dragon, the largest of all lizards. Crocodiles occur in the tropical parts of the region and there are three families of turtles, one of which is unique, one of which occurs also in the Oriental region and one in South America. The turtles shared with the Neotropical region belong to the chelyid family of side-necks. The Australian chelyids have strikingly long necks and are aquatic.

Amphibians are few, Australia being the only region from which common toads are absent. A few common frogs have reached the region and there are hylid tree frogs, found also in the New World and Palearctic, but absent from the Oriental and Ethiopian regions. There are no tailed amphibia.

The freshwater fish fauna is equally poor, but the third lungfish is found in the rivers of Queensland. *Neoceratodus* differs from the other two lungfish in the more obvious development of its lobed fins.

The Australian vertebrate fauna is remarkable for the poverty of its fresh-water fish, amphibia and reptiles and for the uniqueness of its mammals. Some part of the fauna, a few frogs, turtles and marsupials, resembles that of South America. But another part, made up of the terrestrial reptiles, many birds and the placental mammals, shows close affinities with the Oriental region. The Australian region has little in common with the Ethiopian, for although they both have lungfish and side-necks, they do not share the same families of either of these groups.

4

Dispersal

Creation

Before Darwin and Wallace announced the theory of natural selection it was generally assumed that each species lived in the region best suited to it because it had been especially created for that place. Sloths were created for South America, elephants for Africa and India, and rats presumably for the whole world. Creation determined the location and the number of any group of animals. Creation explained both continuous and discontinuous distribution.

Certain facts however do not fit the theory of creation satisfactorily. The rapid spread of the rabbit in Australia after its introduction by man has proved that there was no obvious reason why the rabbit should not have lived there before, if congenial surroundings were all that was required. House sparrows have spread widely through North and South America; frogs which were once absent from the Azores attained plague proportions; the grey squirrel of America has proved more successful than the native British red squirrel and is slowly supplanting it.

Creation does not provide an answer either to the question why animals are limited or why they spread.

Limitation

There are four main causes of limitation of range on land, three of which are closely interdependent. The first three are climate, vegetation, and other animals; the fourth, physical barriers. An animal may be limited by any one of these causes, by a physical barrier for instance, or it may be and usually is, limited by more than one. The causes of limitation form a complex network round the animal. The way in which these factors interact are discussed in detail in Allee and others (1949), Hesse, Allee and Schmidt (1951) and Elton (1958). Here limiting factors are only outlined briefly.

Climate can be thought of as composed of temperature and rainfall.

Low temperature prevents animals adapted to tropical conditions spreading to the poles. Reptiles which are primarily tropical, being numerous and varied where temperatures are high, decrease both in number and variety towards the poles. Crocodiles are hardly known outside the tropics and the most northerly occurrence of a turtle, the European pond tortoise, is $57\frac{1}{2}°$, and this is exceptional for the order. Conversely, high temperature prevents animals adapted to a cold climate spreading to the tropics. Dalliid blackfishes are limited in this way to northern polar regions. Penguins, too, are limited to cold water but, because cold currents flow north from the Antarctic, they have in one case reached the equator.

On mountains, temperature falls with increase in height. This can restrict the passage of an animal over a mountain range or, conversely, can restrict the spread of a high-mountain animal through the warmer lowlands. Parrots are rarely found in mountain regions where temperatures are low.

The other component of climate, rainfall, has a less obvious effect on animal spread. Vegetation is very sensitive to changes in rainfall and it is through its effects on the plant life of the land that rainfall is principally instrumental in limiting the spread of animals. Even so, most animals are prevented from living in or crossing deserts as much because they cannot survive long periods without water as because of the scarcity of vegetation. Animals that are specialized for living in deserts often have exaggerated powers of water retention in their tissues. At the other end of the scale, amphibia are more numerous in wet parts of the world than in dry because the majority are dependent on water for reproduction and their outer skins are not impermeable.

CONIFEROUS
FOREST AND
DISTRIBUTION OF
THE SCOTTISH
CRESTED TIT

The distribution of the vegetation of the world depends in a large degree on the climate, on both the temperature and the rainfall. Some animals can survive in many types of vegetation but the majority are limited. Tapirs and vicunas are both mountain-living animals, but the tapirs frequent the forests, vicunas open plateau land. The European crested tit nests in coniferous forests and its range in the British Isles is therefore limited to the central highland area of Scotland. The koala bear lives in the Australian region where there are eucalyptus leaves to eat, and the giant panda in China where there are bamboo shoots.

The third factor in animal limitation is the interaction between all the animals themselves. This is a complex problem. Animals prey on one another, animals parasitize one another, animals compete with one another for food, nesting sites, or just for space to live. When a new animal moves into an area it upsets the balance of the native population. If the population is sparse the newcomer may be able to move in with little upset. If the newcomer is a carnivore it may destroy much of the old population, as the fox is in danger of doing in Australia. In other cases, an animal which neither preys on nor parasitizes any member of the population may move in and oust one or more of the original population. This seems to happen when two animals have identical requirements for food, shelter and reproduction, and is known as replacement. The increasing limitation in range of the British red squirrel before the advances of the American grey squirrel may be a case of replacement. Amongst carnivores, the dingo dog may have replaced the Tasmanian wolf in Australia.

The fourth limiting factor is the presence of physical barriers. On land these barriers may be rivers, mountains or deserts. The last two of these differ from the first in that the limitation may be due to climate and vegetational factors more than to the actual physical barrier. Climate and scarcity of vegetation are probably more important in preventing animals crossing deserts than, for instance, the inability to walk on sand or stones. Mountains may be too high for an animal to cross but again it is more likely to be climate and vegetation that are limiting factors. Water is the least complex physical barrier. The river Amazon provides a barrier for many animals including monkeys and butterflies. But rivers are not a serious barrier to the average land vertebrate and they form a ramifying highway for freshwater fish. It can be assumed, therefore, that vertebrates spread across land unless they are prevented by one of the terrestrial limiting factors. Limitation by these factors is of major importance in determining the extent of range of an animal within a continent.

Spread

Land vertebrates spread across land if conditions are favourable, but even if conditions are ideal some animals will spread faster than others. It may be a matter of size, psychology, the relative rates of birth and death, or it may depend simply on locomotion. Flying vertebrates should disperse faster and further than ground-living forms, but there are so many limiting factors operating everywhere, that even their potential spread is considerably reduced.

If powers of dispersal on land are not normally important limiting factors for vertebrates, differential powers of dispersal across water lead to many apparent anomalies of distribution. Fresh water limits but it is often small in extent. Therefore the main obstacle to wide spread is sea water.

Freshwater fish are variable in their response to sea water. Some, like the lungfish and perch, are unable to tolerate salinity for more than a short time, others, like salmon and eels, are as much at home in sea as in fresh water. For these last, the crossing of the sea is not a serious problem. These two extremes grade into one another. There are some minnows that are moderately salt-tolerant and could probably survive a sea journey, and sticklebacks appear to tolerate fresh and salt water equally.

Amphibia can swim, but they cannot tolerate salt water. Only one species of *Rana* is known to enter salt water normally and survive. Their permeable skins, and their reproductive habits are adapted to life in fresh water or damp terrestrial surroundings. Amphibia could not swim through the sea. Some might be carried through the air by birds, but the same characteristics that make them intolerant of sea water make them susceptible to drying out in the air. They cannot swim in the sea and they cannot be exposed for long periods to dry air. The only possible method for them to cross the sea seems to be provided on rare occasions when what is virtually a small island is swept into

the sea by exceptionally heavy rains and winds. In general then, amphibia do not normally spread across salt water barriers. Only on rare occasions do they seem to have 'rafted' a short distance. Although rafts have often been sighted, no one has studied the fauna of a raft, its place of origin nor its subsequent arrival. Driftwood from distant shores has been found washed up in some parts of the world, on some occasions with wood-boring animals in it, but no one has been able to study the whole process, and rafts must, therefore, remain a hypothetical means of transport.

Reptiles are much more versatile. The physiological limitations of amphibians do not apply to reptiles. It has been claimed that land and fresh water turtles when closed tightly inside their carapace, can float unharmed in the sea for long periods, and certainly they have colonized many islands hundreds of miles out in the ocean. Adult reptiles of all sorts and their eggs may be swept across stretches of sea on natural rafts. As neither the adults nor their eggs are as susceptible to drying as the amphibians, the rafts or islets may not need to be as large and leafy as those on which amphibians are supposed to travel, and they may, therefore, occur more often. Sea water is not, therefore, an insuperable obstacle to reptiles.

Birds because they can fly are able to overcome many of the barriers to dispersal which prevent other animals from spreading. Theoretically there is no limit to the spread of birds, but some birds fly further afield than others for reasons which are not always clear. Many birds are prevented from spreading for 'psychological' reasons, and the more usual limiting factors are as important in bird dispersal as in other cases. Furthermore, just because birds can fly, they can fly back to their starting point.

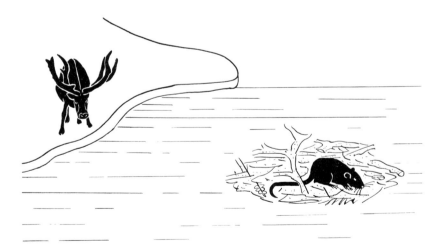

Land mammals vary in their response to sea water. The amphibious hippos and otters for instance are good swimmers and can probably survive long periods in the sea. Pigs, too, swim well and far. Bats, like birds, can fly over the oceans. Rats seem to be the most accomplished of all rafters if their world-wide dispersal and universal presence on ships is taken as evidence.

Spread across the seas is easier for some land vertebrates than it is for others. Amphibians are the most confined to land. Mammals are nearly as restricted, with the exception of bats and rats. Fresh water fish vary greatly amongst themselves. Birds and reptiles are the least confined.

It has been assumed that animals are always tending to spread and that they go on spreading until the limiting factors become intense enough to stop them. Having stopped they have been presumed to keep the range they had attained. This is not always true. The range of an animal can contract. The range of the red squirrel is gradually contracting before the spread of the grey.

Regional Aspects

Land vertebrates spread across land and with varying success across the sea. The complex interaction of factors which limit this spread goes some way to explain the faunal differences between zoogeographical regions. Climatic conditions may be the main reason why urodele amphibians are absent from the southern continents but widespread through the north temperate regions. The vegetation may be the main reason for the small range of the giant panda. The success of flightless rails in New Zealand has been due mainly to the absence from that island of carnivorous mammals. Sea water accounts for the absence from the Azores of native frogs.

Other differences between the regions are not so easily accounted for. It is not clear on these grounds alone why tapirs should be absent from the Ethiopian region. Inability to spread in itself would not seem to be an explanation, since they have somehow spread between South America and Malaya. Climate and vegetation do not seem to explain the distribution either. Climatic and vegetational features similar to those in which they live can almost certainly be found in Africa. Again, these factors do not seem

adequately to account for the presence of marsupials in Australia and the Americas and their absence from all other regions.

The same sort of problems arise when the faunas of characteristic vegetational or climatic zones in different parts of the world are studied. Although the conditions of life may be almost identical, the fauna is unlikely to be so. The fauna of the great deserts of North America and Africa though superficially similar is not identical. The kangaroo rats of Arizona and the jerboas of the Sahara are superficially alike. Their hind legs are long, the feet furred, tails long and tufted and their hair silky. Close anatomical inspection reveals however, that the structure of their skulls is different, the kangaroo rats, therefore, belong to the family Heteromyidae, and the jerboas to the family Dipodidae. Similarly the horned lizards of Arizona, though strikingly similar to the spiny-tailed lizards of the Sahara, belong to separate reptile families, to the Iguanidae and to the Cordylidae. There are antelopes in the African desert, Gila monsters in the American. The differences are extensive but not complete. The vipers of the Sahara and the rattlesnakes of North America both belong to the same snake family, the Viperidae. The ground-squirrels of the Sahara and the prairie-dogs of the Nearctic desert are both Sciuridae.

Animals then do not live wherever climatic and vegetational conditions are suitable for them, nor are they only absent from land surrounded by salt water. Although some of the faunistic differences between the zoogeographical regions can be explained in these terms, others seem to require a different approach to the problem. The vertebrate fauna of the regions has been described as it is today; the means of dispersal and the limiting factors are those which are operative today. But today is an arbitrary moment. Behind it in time stretch millions of years of animal life, and behind that many more millions since the beginning of the earth. At all times animals have spread and been prevented from spreading, but they have not always been the same animals as today, nor have the limiting factors had the same spatial distribution. What has been happening over the years has affected the patterns of distribution that are seen today. An explanation of the anomalies of today must be sought in the conditions of the past.

Part Two
PAST

5

Paleogeography

Rocks

The earth is very old. It had its beginning some 4,000 million years ago. In its beginning it was a whirling mass of matter which gradually condensed into a sphere, and a crust was formed. As the surface of the crust cooled, water vapour from the cloud of gases surrounding the planet condensed to give rise to the oceans. It is not known how the land and sea were distributed at this early time. The oldest rocks are 2,000 million years old, and they are found on every continent, but their extent is impossible to estimate.

The land rose and fell many times during the succeeding years, mountains were thrown up and were worn away by the wind and the rain, the ocean invaded new areas and left others high and dry so that the face of the earth was changed considerably. But in spite of all this rising and falling, crumpling

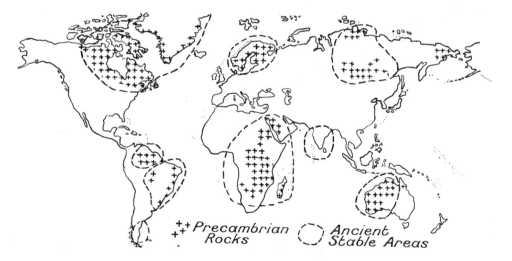

++ Precambrian Rocks   ◯ Ancient Stable Areas

and faulting, it is possible that parts of the main continents had some sort of existence through this time. Furthermore, the presence of the oldest rock formation, the precambrian shields, on each of the six continents, suggests that none is much newer or older than any other. Unfortunately it is not known whether rocks of similar age and characteristics lie on the bottom of the oceans, the remains of lost continents.

Coming to epochs less remote in geological time, more precise statements can sometimes be made. Sediments accumulating on a sea floor or lake bottom are the source of rocks which are the most useful in the interpretation of past changes in the land. The sediments may be of fine sand or silt or the shells and skeletons of aquatic plants and animals. In time the lower layers are compacted by the pressure of the layers above and form into hard rock. An upheaval of the sea bottom, the draining away of water from a lake or a tilting of the land, may uncover such sedimentary rocks and bring them to the surface. Once above the surface, the rock is subject to the forces of erosion and thus succeeding layers become exposed. The nature of the inorganic sediments which have built up the rock may provide a clue to a former coastline, river bed or delta. But mainly the clues in the rocks are sought from the animals and plants that went to make them. Deposits of marine animals in a layer of rock indicate that the particular area was below the sea at the time when the animals were alive. Remnants of land plants or land animals can indicate dry land, but they do not always do so. An animal is rarely preserved unless it falls into water and although a land animal may remain in the fresh water into which it first fell it is frequently swept by rivers away from the place where it lived and even into the sea before it is fossilized.

The succession of plants and animals in stratified rocks determines the relative age of the strata. The presence of closely similar animals and plants in rocks in different parts of the world is taken as evidence of the contemporaneity of the strata. As a result of paleontological studies, and studies of different types of rock, the following time scale has been drawn up, dating from some 600 million years ago when the rocks contain for the first time a number of recognizable animal fossils.

GEOLOGICAL TIME SCALE, IN MILLIONS OF YEARS SINCE THE BEGINNING OF EACH EPOCH
(Holmes 1960)

	0 recent	
	1 pleistocene	first fossils of man
	10 pliocene	
	25 miocene	
Cenozoic	40 oligocene	
	60 eocene	
	70 paleocene	
	135 cretaceous	
Mesozoic	180 jurassic	first bird fossils
	225 triassic	first mammal fossils
	270 permian	
	350 carboniferous	first reptile fossils
Paleozoic	400 devonian	first amphibian fossils
	440 silurian	
	500 ordovician	first vertebrate fossils
	600 cambrian	

Both plants and animals, invertebrates and vertebrates are used in such studies. Obviously some animals are more likely to be fossilized than others, some more easily identified, land animals more useful for determining former land, marine animals for former seas. For remote epochs the geologist or the paleogeographer relies on the well-preserved invertebrates for his evidence. The skeletons, shells, of molluscs for instance, may be preserved intact in the rocks, or more frequently the substance of the living organism may be replaced by mineral silica or calcite, and leave a cast or impression of itself. Ancient invertebrates can show that the limestone rocks running across southern England in a narrow band from the Dorset coast to Lincolnshire were below the sea some 150 million years ago.

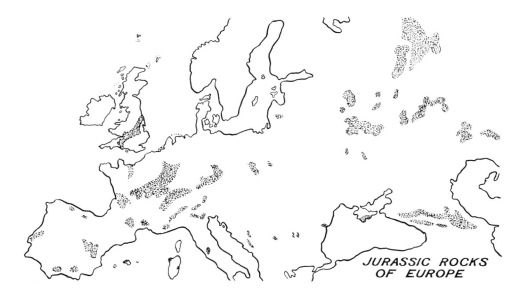

JURASSIC ROCKS
OF EUROPE

To locate former sea on present-day land masses is not then very difficult. To identify the extent of the land in the past is less easy. Explorations of the sea bottom are in their early stages. Land animals are not preserved in exactly the place where they lived, and they are not very likely to be preserved at all. If they are fragile their skeletons are crushed or disintegrate before reaching the sedimenting waters of lake or sea, if they lie on land which is acid they are eroded. Such factors as these have led to a scarcity of insect fossils. Those preserved in the resin of Baltic coniferous trees date only from

30 million years ago. Amphibians are fragile and very few Mesozoic amphibians are known. The skeletons of birds too are easily destroyed before fossilization can take place and they are also difficult to piece together when they are found. The most useful fossils for identifying ancient land masses are those of terrestrial reptiles and mammals. This sets a limit in time below which it is unprofitable to go: a time before either reptiles or mammals were abundant on the land. But because reptiles cross sea barriers more easily than mammals, they give less useful evidence of land continuity in the past. In the present state of knowledge, world paleogeography depends to a large extent on a knowledge of mammal fossils.

There is another aspect which affects the reliability of fossil distribution as a guide to former land masses. This is the chanciness of finding fossils. Most are found by accident and depend on excavations that are carried out for other purposes, railway, road and factory building. Their density is therefore to some degree a measure of the industrial development of a country. Other fossils are found because paleontologists go to look for them. These tend to measure the distribution of paleontologists as much as that of the fossils themselves.

Positive evidence provided by a particular fossil in a particular place is valid, but, for the reasons given, negative evidence, that no fossil has been found in a particular place, is open to doubt. That fossil could turn up some time.

This is the sort of evidence on which maps of the ancient world have been built up. Considering the fragmentary nature of the evidence it is not surprising that there are many versions of these maps. There are also several different theories to account for the supposed past distribution of the land. The geographer can discuss their relative merits in terms of present-day geography, but the zoogeographer must argue from prejudiced evidence. When he tries to account for past and present distribution in terms of past land patterns he is arguing in a circle if those land patterns have been surmised largely from a study of past and present distribution. The zoogeographer needs independent physical evidence of the state of the continents through the ages to fit his plants and animals to, but unfortunately such independent evidence is usually not available. Each of the best known theories about the land in the past (from whatever arguments they have been derived) must therefore be considered. The decision to adopt one or the other will depend on which fits the facts the best and which is the simplest, always remembering that the theory is only a guess and that the facts at present available are not final.

The three best-known theories of paleogeography are those of the permanence of the continents, land bridges, and continental drift.

Because it is the simplest of the theories and because it fits a great many of the facts adequately, the first of these theories will be assumed to be the most likely, and the biological consequences of accepting it considered. This does not, however, mean that it is the correct theory but only that it forms a useful model to work with. The other two theories will be discussed later in relationship to the biological findings of the first.

Continents

The theory of the permanence of the continents was put forward by Lyell in 1830–33. He argued that the great continents have always had some sort of existence and have occupied the same relative positions as they do today, but he also argued that these continents have changed in detail, have evolved through the geological ages as a result of the same sort of physical forces that are at work today. On the basis of this theory, world paleogeographical maps have been made. Some stretch back into cambrian days (see Wills 1951 and Dunbar 1949) but these are usually confined to limited areas. Maps of the whole world in these remote times are too indeterminate to be of much help to zoogeography. They take on more definite shape only in the middle of the Mesozoic.

Some 180 million years ago some part of all the continents and the main

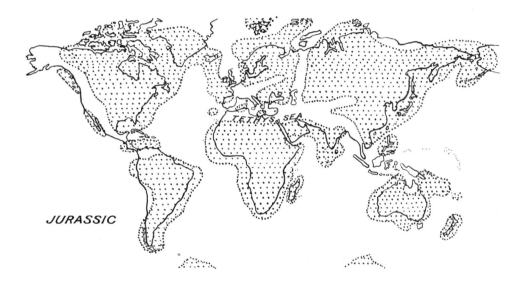

JURASSIC

oceans are recognizable but they are not joined up in the same way as they are today, and they are only roughly the same shape. The most obvious difference between then and now is that the northern and the southern land masses were almost entirely separated from one another by a vast sea, the Tethys Sea. The northerly and westerly parts of the British Isles formed part of a North America-Greenland continent. Most of what is now Europe was below the Tethys Sea.

Several lines of evidence then suggest that by the cretaceous, South America and Africa had gained connexions with the northern continents by corridors of dry land. These connexions were of comparatively brief duration. Africa may have been isolated again before the end of the cretaceous, and South America only a few million years later, losing its North American connexion in the early paleocene. For the rest of the paleocene and eocene, neither of these continents had any connexion with the northern land masses.

Africa was the first to regain contact with the north, possibly in oligocene days, and only much later, in the pliocene, did the isthmus of Panama once again connect South America with North America.

During this later period when the African and South American connexions were being made with the north, the British Isles became part of Europe and ceased to be an outlying strip of the America-Greenland continent.

During the Cenozoic, too, changes took place along the shallow sea area between Alaska and Siberia. Several times there seems to have been land across what are now the Bering Straits and at some times this land may have included the more southerly islands of the area. The land was not stable in this part of the world however, and was as many times below the sea.

At different periods during the history of the earth there have been times of great movements when mountain ranges have been thrown up. The Caledonian mountains were formed as early as the silurian, the Appalachians of North America in the permian, the Andes and Rockies in the late cretaceous. The Eurasian mountains are mainly young. The Himalayas which cut off the Oriental region from the rest of Asia were formed as late as the middle of the Cenozoic, at the same time as the Alps and Pyrenees.

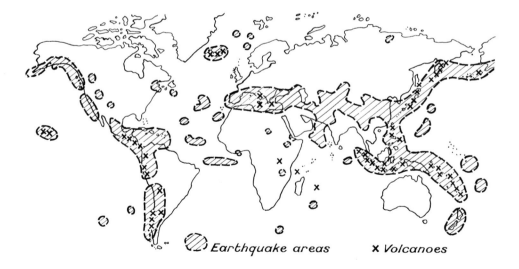

Earthquake areas   × Volcanoes

All this time too there was volcanic activity over parts of the world, probably even more widespread than it is today. An area that is volcanic still, has probably been an unstable part of the world through many epochs. Melanesia and the Malay Archipelago northwards through Japan and the Aleutian Islands have long been changeable owing to volcanic movement. There are volcanic areas also in the west of Central America and South America, and again in the Mediterranean.

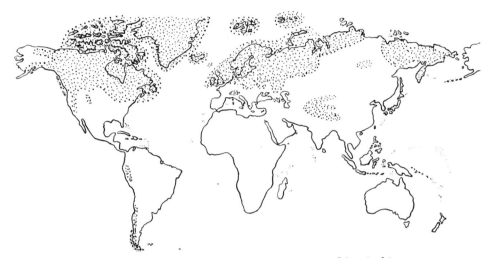

Maximum extent of Pleistocene Glaciations

In the last epoch of the Cenozoic yet another physical event had a profound effect on the world. Less than a million years ago, in the pleistocene, great sheets of ice covered areas of the northern hemisphere, advancing and retreating several times. Ice-covered areas were uninhabitable except for a very few species of animals and plants, and south of the ice itself the climate was cold. The general effect of the ice was to lower the temperature all over the world, pushing the edge of the temperate zone further south and reducing the extent of the tropical areas of the world. The accumulation of ice, lowered the level of the sea so that many areas that are now shallow seas were dry land during the pleistocene.

Over the past 180 million years, then, according to the theory of permanence, the main changes in the look of the world had been the shrinking of the Tethys Sea, the making and breaking of narrow connexions between the two northern continents and of their connexions with South America and Africa. Australia had been isolated most of this time.

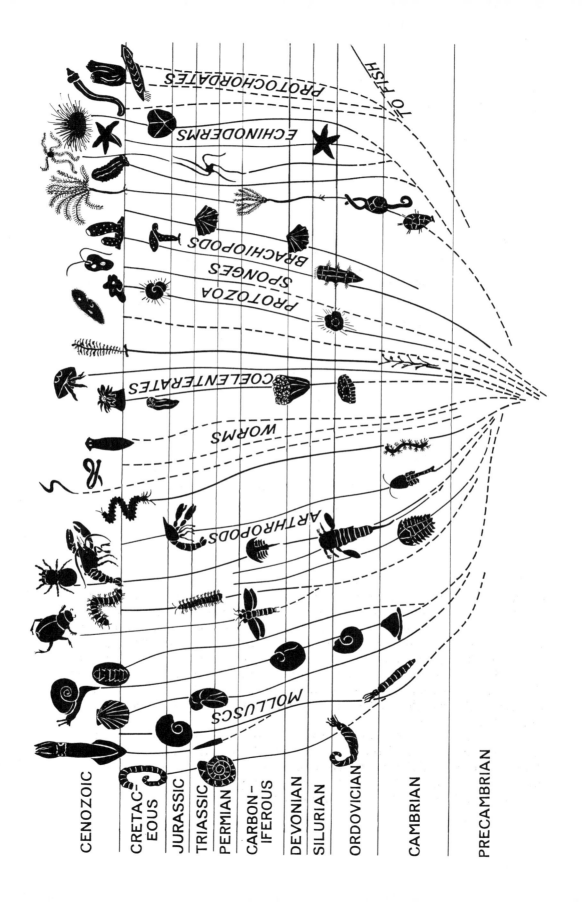

PROTOCHORDATES

TO FISH

ECHINODERMS

BRACHIOPODS

PROTOZOA

SPONGES

COELENTERATES

WORMS

ARTHROPODS

MOLLUSCS

CENOZOIC

CRETAC-
EOUS

JURASSIC

TRIASSIC

PERMIAN

CARBON-
IFEROUS

DEVONIAN

SILURIAN

ORDOVICIAN

CAMBRIAN

PRECAMBRIAN

6

Evolution of Animals

Paleozoic and Mesozoic

Living organisms have a long history, extending back possibly over 1,500 million years. It is not known what the first forms of life were, nor when plants and animals diverged from the common stock. It is presumed that the first recognizable animals resembled the Protozoa of today. The early forms may have been like the amoeba of the rain-water puddles, or like the plant-animals that move by the lashing of their whip-like flagella. After these came the sponges, coelenterates (jellyfish, corals and sea anemones) and flatworms. Gradually other forms of life came into being, other worms, molluscs (bivalves, snails and cephalopods), arthropods (crustacea, spiders, insects) and echinoderms (starfish, sea urchins, sea cucumbers). Almost nothing is known of the early evolutionary history of these invertebrate groups, for they had already a long history behind them by the cambrian. Cambrian is the name given to the earliest rocks in which recognizable fossils are normally found. The main groups of invertebrates are found in these rocks but they are represented by animals very different from modern forms. The most abundant cambrian animals are the trilobites, arthropods which looked like aquatic woodlice. They all lived in the sea, mostly on the bottom. There were no land animals in the cambrian. Trilobites lived on, becoming less abundant but not dying out until the permian. Besides worms, coelenterates like *Obelia*, and jellyfish, there were in cambrian seas eurypterids and king crabs, both relatives of modern scorpions. The extensively armoured eurypterids, some measuring six feet, dominated the seas and fresh waters as the trilobites declined. There were several sorts of molluscs, snail-like forms whose shells were not twisted, early bivalves, and shelled cephalopod ancestors of the nautilus and modern octopuses. All cambrian echinoderms were fixed to rocks by stalks and were covered over by thick calcified skeletons. According to the fossil record, in the early years of the Paleozoic, heavy arm-ouring was the rule, but it should be remembered that forms with hard skeletons are more likely to have been preserved than those without so that the balance would be in favour of armoured forms. Even so, when the first known vertebrates appeared in the ordovician they too were heavily arm-oured.

The first vertebrates, fish without jaws, lived in the fresh waters of the period. Later still, life began to colonize the land. In the devonian the earliest known land plants, club-mosses, ferns and seed ferns came into being, and

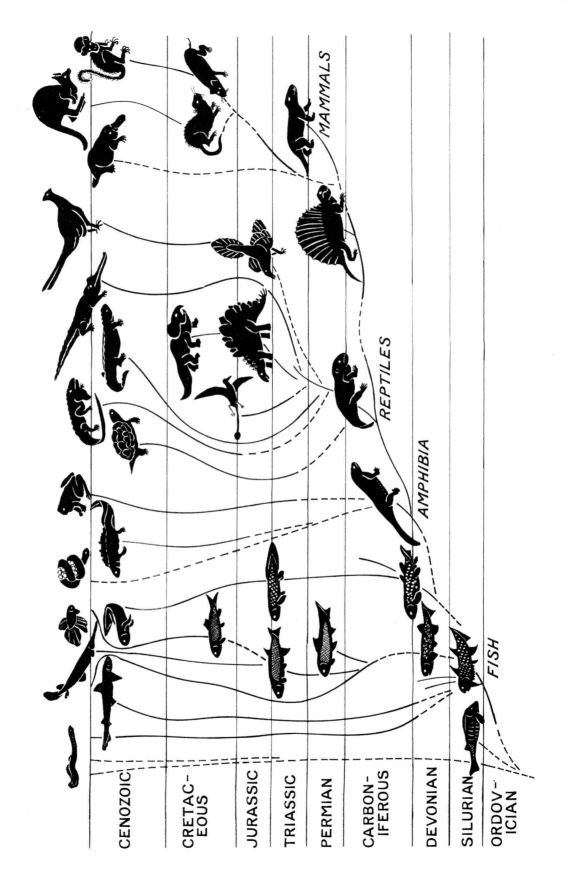

CENOZOIC

CRETAC-
EOUS

JURASSIC

TRIASSIC

PERMIAN

CARBON-
IFEROUS

DEVONIAN

SILURIAN

ORDOV-
ICIAN

MAMMALS

REPTILES

AMPHIBIA

FISH

shortly afterwards the earliest scorpions, spiders, centipedes and insects. The insects had not yet evolved into flying forms. Fish had become abundant by this time, radiating out into different forms, many of them still heavily armoured. The devonian forms died out eventually, but not before the ancestors of all modern fish had been derived from them. As the cambrian was the Age of Trilobites, the ordovician and silurian the Age of Eurypterids, so the devonian can be called the Age of Fish.

In the next geological epoch, the carboniferous, the vertebrates came out on to the land in the form of amphibia and reptiles. Insects were numerous by now and the carboniferous has been called the Age of Cockroaches. For some time to come, however, the land was to be dominated by the reptiles which evolved into many strange forms. There were small reptiles, large reptiles, quadrupedal reptiles, bipedal reptiles, herbivorous reptiles, carnivorous reptiles, land-living reptiles, aquatic reptiles and flying reptiles. The reptiles reached their peak in the jurassic and then lost their superiority. The abundant dinosaurs both large and small, and the flying reptiles died out suddenly in the cretaceous. It is not clear why this happened and many different theories have been put forward to explain it. It may have been too cold, or too dry, the vegetation may have become unsuitable as food, or some disastrous disease may have killed them or their eggs. It was most probable that it was some change in the climate of the cretaceous which was responsible because other groups of plants and animals were seriously affected at this time. Many of the cycad-like plants and the ginkgos died out on the land, and from the sea the big marine reptiles and the ammonites and belemnites (two groups of cephalopod mollusc) disappeared.

Whatever the cause of the reptile catastrophe, the birds and mammals were able to take advantage of it. Both these groups, the only two that are warm-blooded, were evolving whilst the reptiles were still the most successful land animals. By the cretaceous, birds and several groups of early mammals were in existence. They seem to have found suitable conditions for their way of life at the end of this period possibly provided by the newly evolving flowering plants and higher insects, for they diverged rapidly and started evolving into all the groups that are known today. Mammals, birds and insects have dominated the land since the beginning of the Cenozoic.

The Paleozoic and Mesozoic record is the story of succeeding groups rising to dominance and gradually dying away, from the Age of Trilobites through the Ages of Fish and Cockroaches to the Age of Reptiles. The Cenozoic is the Age of Mammals.

Cenozoic

By the jurassic, then, most of the main groups of animals had already a considerable evolutionary history behind them. The five classes of vertebrates, the fish, amphibia, reptiles, birds and mammals had already been established.

Since the jurassic no new orders of fish, amphibia or reptiles have come into being, but within the orders already established there has been divergence into the families which survive today. Birds and mammals have diverged into families as well as orders since the jurassic. Families were used as the units in describing geographical distribution, so to be consistent it is also the family that must be considered as the evolutionary unit.

Ideally all modern families could be traced in detail from their first appearance in the rocks to the present day. This is possible for some mammal and a few reptile families. For the others, adequate fossil evidence is still lacking.

The main fish orders were already well established long before the beginning of the Cenozoic, and in fact many of the modern families of freshwater fish were already differentiated by the cretaceous. For instance, ancestors of the Australian lungfish were already similar to their modern descendants by this time. Holostean ancestors of the Nearctic garpike and bowfin, and the first characin fish were also marked off by the cretaceous. The carp family was a later development, not appearing until the eocene.

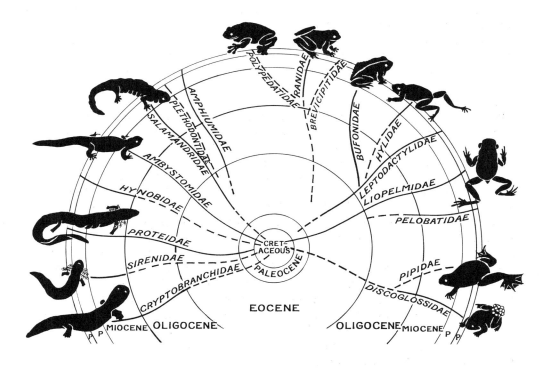

Almost nothing is known about the history of the modern amphibia. The three orders, apoda, urodela and anura are difficult to trace and difficult to relate to one another because of the scarcity of fossils. When the first anura appear in the rocks they are already well differentiated from other amphibians and show little signs of their derivation. It is therefore from the comparative anatomy of modern families that a hypothetical scheme of their Cenozoic evolution has been worked out. On this basis it seems that the liopelmids

(*Liopelma* and *Ascaphus*) were the first of the modern anurans, in the late jurassic, followed by pipid toads (*Pipa* and *Xenopus*) and, in the oligocene, hylid tree frogs and the common toads. The differentiation of the common frogs and polypedatid tree frogs may have been as late as the miocene. The ancestry of the urodeles is equally obscure. The ambystomid family (axolotls) is found in the cretaceous, salamanders in the eocene and giant salamanders in the oligocene. Their relationships to one another are not easy to determine. Nothing can be said about the evolution of the apodans.

Of the reptiles, only four orders persisted through the Cenozoic after the great Mesozoic orders had become extinct. These were the turtles, crocodiles, snakes and lizards, and *Sphenodon*. The turtles were recognizable by the triassic and during the jurassic differentiated into the two groups that are known today: the pleurodires (side-necks) and the cryptodires (vertical-neck turtles). Both groups had split up into their modern families by the cretaceous. Crocodile families too were firmly established before the beginning of the Cenozoic and this is also probably true of lizard and snake families. The varanid and iguanid families may be the oldest, and both they and the aga-mids (possibly derivatives of the iguanids) are known from the cretaceous. Teiids, lacertids and chameleons may be later offshoots, together with the snakes which almost certainly derive from varanid-like lizards. The *Spheno-don* family is the oldest of the surviving reptiles being already abundant in the early Mesozoic.

Birds left the main reptile stock, probably in the jurassic, and continued to evolve along independent lines, leaving little trace of their subsequent history in the rocks.

It was for the mammals that the Cenozoic was so important. It was their great evolutionary moment. Before this mammals had been scarce but in a comparatively short time they became the dominant land animals.

In many cases their fossil history is well known and some families can be traced without a break from their first appearance to the present day. But it is just this completeness which itself raises a difficulty. In theory all mammals could be traced back along converging lines to one common ancestor amongst the reptiles of early Mesozoic days. In theory the lines should be continuous from the earliest forms to the most modern. This hardly ever happens, how-ever, and groups are separated from one another because their ancestors are not known. In practice, therefore, it is possible to talk about the duckbilled platypus family because it is cut off from its nearest relative the spiny anteater by certain morphological characteristics and there are no fossils earlier than the pleistocene to make the two families merge into one another. It is a good deal more difficult to decide when an elephant becomes an elephant. Ances-tors of all elephants, mastodons and mammoths are known in the eocene and an evolutionary tree can be drawn giving off families of mastodons and the elephant family by continuous branching. Who is to decide in such cases when the elephant family has started, when it has ceased to be a pre-elephant or a paleomastodon and become an elephant? Unfortunately no word exists to describe this process in terms of classification and the decision to stop calling the fossils paleomastodons and to start calling them elephants at a particular moment in geological time is an arbitrary one. The better known the fossil history of a family the more acute is this problem. Bearing in mind

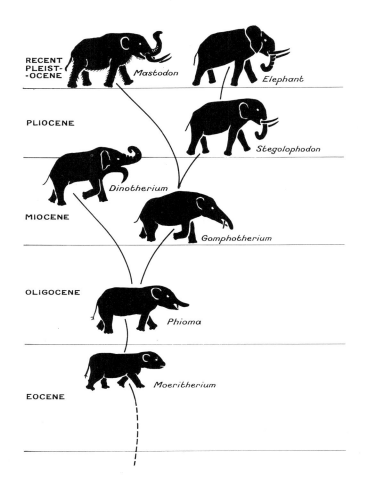

this arbitrariness of saying when a particular family has started, an outline of cenozoic mammalian evolution can be traced.

The first marsupials and first placentals appear in the rocks at the same time, in the cretaceous, giving no indication of their relationships to one another. From then on the fossil record is good.

Fossils of the main orders appear in the following sequence: marsupials and insectivores, edentates, condylarths and carnivores, primates, lago-morphs and rodents, followed by bats, perissodactyls, artiodactyls and proboscids.

Thus the Insectivora were the first of all placentals, differentiated in the cretaceous and contemporary with the first marsupials. They were ancestral to the later placental orders. Of insectivore families, hedgehogs and moles were the earliest fossil representatives of modern families.

No other mammal orders are known as early as the cretaceous.

In the next stratum however, the paleocene, several orders are found, distinct from the insectivores from which they were derived and distinct from one another. These are edentates, condylarths, carnivores, primates, lagomorphs and rodents. Of these the condylarths and primitive carnivores, creodonts, are more nearly alike at this time than the others, and show clear signs of derivation from the insectivores. They were both clawed animals with long tails, but whilst creodonts were probably still mainly insect-eaters, the condylarths were herbivores. From these generalized paleocene ancestors have evolved modern carnivores and the hoofed (ungulate) herbivores of the world. Condylarths became extinct during the eocene, unless the modern aardvark be considered a member of the order.

Creodonts, the earliest of the carnivores, radiated widely until they were replaced by modern carnivore families in the oligocene. By this time dog, civet, weasel and cat families were clearly distinct, the cat family having already sabre-tooth tigers amongst its members. Later came the raccoon family; and the bear family had differentiated from the dogs by the miocene. These were followed by hyenas which were the last of all the carnivore families to evolve, being derived as late as the pliocene from the civet family.

Of the other paleocene orders, edentates were represented by the paleanodonts, early derivatives of the insectivore stock. They gave rise rapidly to armadillos, and much later to anteaters and sloths.

Primates, whose affinities with the insectivores were very close in the paleocene, became common in the eocene and by the oligocene both Old World monkeys and apes had become distinguishable, to be followed closely by the New World monkeys and lorises.

The remaining two paleocene orders, the Lagomorpha and the rodents can be presumed to have arisen, like the others from the insectivores; but whether they were independent derivations from the start or whether they shared a common ancestor in late cretaceous days is not known. The Lagomorpha were at first rare, expanding in the eocene but not becoming really common until the oligocene by which time both the rabbit and pika families had differentiated.

The story of the rodents is more difficult to follow, for the relationships of even the modern families have not been satisfactorily worked out. They occur as rare fossils (probably sciuromorphs) in the late paleocene, expand in the eocene when ancestors of the sewellel family were already found, and by the oligocene the main groups had been sorted out. The sciuromorphs were giving rise to the beaver family, the myomorphs to the cricetid mice and gophers, and the caviomorphs to tree porcupine, cavy and chinchilla families. The squirrel family itself did not appear until the miocene to be followed, as late as the pliocene, by murid rat and mouse family and the spalacid mole rats.

Perissodactyla and Artiodactyla were in existence in the early eocene

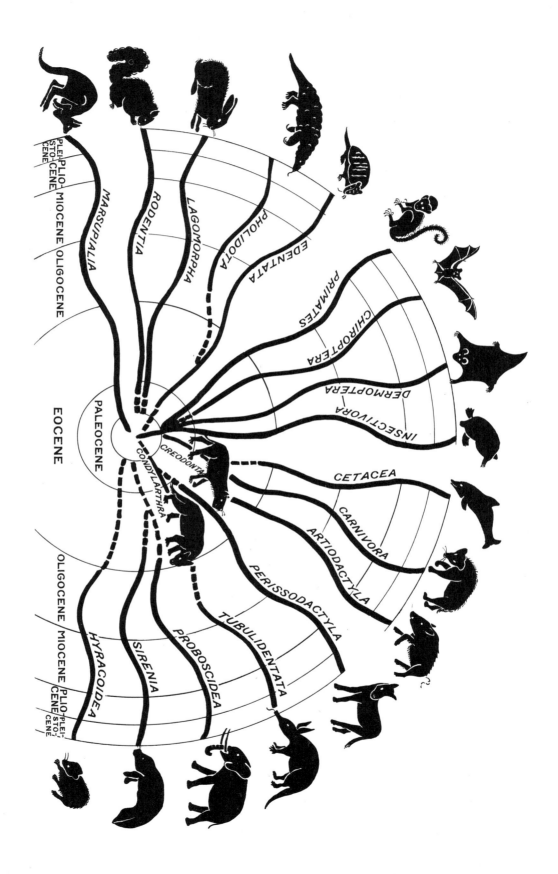

derived from the condylarth-creodont stock. Already the horse, tapir and rhinoceros families were distinct. Of the artiodactyls, the camel family was one of the first to evolve, in the eocene, followed by pigs, peccaries and deer in the oligocene. The miocene saw giraffes, pronghorns and bovids, but the hippo family has not been traced further back than the pliocene although this cannot be the whole story as its differentiation from the pigs must have taken place much earlier. Amongst the bovid family, the antelopes appeared later than the rest.

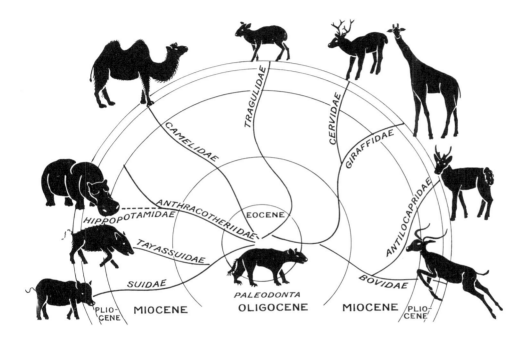

Finally the elephants, whose ancestors were of eocene and possibly condylarth origin, were not differentiated as a definite family distinct from the mastodons until the pliocene.

By the pliocene all the modern families of mammals had evolved, from the early insectivores to the latest of all, the murid mice and the hyenas.

Summary of Mammal Evolution

Cretaceous: earliest mammal orders: marsupials and insectivores.

Paleocene: a few more orders established: edentates, carnivores (creodonts), condylarths, primates, lagomorphs (rare) and rodents (rare).

Eocene: all the main orders established and a few modern families: bats, perissodactyls, artiodactyls (camels), proboscids. Lagomorphs, primates and rodents expand.

Oligocene: a large number of modern families established: Old World monkeys; dog, cat, weasel and civets among the carnivores; beavers, cricetid mice, gophers, porcupines, cavies and chinchillas; sloths; pigs, peccaries and deer.

Miocene: a few late modern families differentiated from the earlier modern families: raccoons, bears, squirrels, giraffes, pronghorns, bovids and anteaters.

Pliocene: the last of the modern families appear: hyenas, murid mice, spalacid mole rats and elephants.

Part Three

PAST TO PRESENT

7

Northern Regions

Cenozoic Distribution

Mammals differentiated and spread during the Cenozoic from small beginnings to today's diversity and range. Fossils of mammal families are abundant enough in the rocks for a reasonable estimate to be made of their range at different geological periods. Thus a distribution map of miocene tapirs can be drawn. This can be compared with the modern distribution map on page 13. In miocene days there were no tapirs in South America, but in contrast to present-day distribution, tapirs lived in North America and Europe.

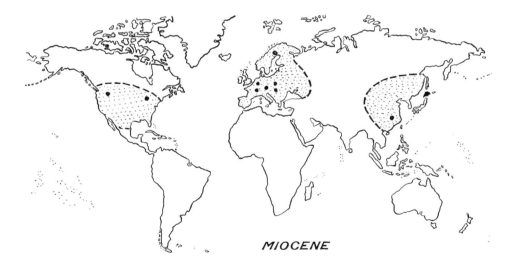

MIOCENE

Just as simple Cenozoic distribution maps for individual families can be drawn, so also can the compound maps of zoogeographical regions of different Cenozoic epochs be delineated. Zoogeographical regions are not modern phenomena; they existed in the past, but they were not necessarily the same as they are today either in continental area or in the composition of their fauna. There are certain unfortunate gaps in parts of the mammal record, but on the whole the changes in range of the main families through the Cenozoic are reasonably well known.

With the theory of the permanence of the continents as a base, a picture can be built up of the evolution of mammal families, their spread, and their colonization of different parts of the world.

The general picture that emerges is one of mainly northern evolution and subsequent spreading from west to east or from north to south. Spread was compensated for by limitation however. Sometimes a family spread widely and then contracted, like the horses, or spread and became extinct in the middle of its range like the tapirs, or like the pronghorns never spread beyond the confines of the continent where it originated.

Whatever the propensities of the families, the direction and the distance of their spread was determined to some extent by the state of the land at the relevant period in Cenozoic history. Their spread was determined by the geographical location of their place of origin and by the connexions that place had with other parts of the world. Thus mammals originating somewhere in north-eastern Asia would first have to spread westwards into mid-Asia or Europe before they were available as migrants to Africa. But they were only able to make this final southerly migration if Africa had a suitable land connexion with the north at the time. The earliest continental connexions therefore provided a migration route for the earliest mammals, the latest connexions for the most modern.

Nearctic and Palearctic

Towards the end of the Mesozoic, the modern Nearctic and Palearctic regions seem to have had a uniform fauna, and for this reason are thought to have comprised one land mass. It is from the northern hemisphere that the earliest mammals are known. By the cretaceous there were both marsupials and insectivores in the continents. There were holostean fish and lungfish; there were crocodiles and side-neck turtles. There were probably also liopelmid frogs, pipid toads and perhaps some of the more advanced frog families spread through the north. But already towards the end of the cretaceous the uniformity of the northern fauna was being modified. Some modern fish and amphibian families were not spreading through the area. Ambystomid urodeles (axolotls) and characin fish are known as fossils from the Nearctic strata of this epoch but not from those of the Palearctic. Marsupials appear to have been more abundant in the Nearctic than the Palearctic. But the differences were not all due to additions in the Nearctic. *Sphenodon* which had earlier been widespread was dying out at the end of the cretaceous and its only northern relic was in the Palearctic.

By the paleocene more orders of mammals had appeared in the north, descendants of the insectivores. There were creodont carnivores, condylarths and lemurs, the first recognizable primates. With the earlier marsupials and insectivores all these orders were spread through the northern lands. But there had also been a loss in the north: lungfish had disappeared. Differences between the Palearctic and Nearctic were becoming more definite. The first lagomorphs were found in Asia. In contrast, the most ancient of all edentates, the paleanodonts, had appeared in the North American continent. There were also a few rodents there. Unlike the rodents and lagomorphs which were to become widespread, the edentates remained

LATE
CRETACEOUS –
EARLY PALEOCENE

in the New World. It is not clear why edentates were restricted in this way, for it is thought that there was plenty of opportunity for faunal exchange between the two northern continents during the paleocene and into the early part of the eocene, judging from the way other groups of animals spread.

Eocene days saw new forms in the northern fauna, and there were no losses of already established groups. The number of mammalian families increased generally. Such is the crudeness of measuring time on the geological scale that several of these families appear simultaneously in both continents. Yet it is unlikely that they could have had such a wide centre of origin. Wherever they originated however, they must have spread rapidly. Such were the horses, rhinoceroses, ancestral artiodactyls and bats. The lagomorphs and

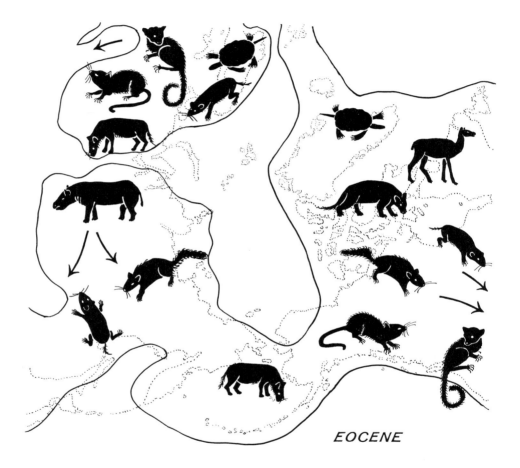

EOCENE

rodents had increased in number and had also spread, the lagomorphs from their original home in the Palearctic and the rodents from the Nearctic. Amongst other vertebrates, characin fish had colonized the Palearctic by the eocene and the holosteans were still flourishing. In contrast to the wide dispersal of these families, even of those which were new to the fauna, others were restricted. They may have been restricted because they were not yet numerous or they may have been restricted because by the late eocene the northern continents had lost their land connexions with one another. Whatever the reasons, there were differences in the late eocene faunas. The Palearctic and Nearctic could be considered as separate zoogeographical regions. The carp and salamander families and a bird, probably ancestral to ostriches, are known from European fossil beds of the epoch. These represented families which were unique to the Palearctic, if only temporarily so. The Nearctic was still characterized by its paleanodont edentates and to these had been added ancestors of the camels.

By the next epoch, the oligocene, further changes had taken place in the fauna. More new families had evolved, but some had also died out from the northern hemisphere. There were no longer any condylarths or lemurs. The other ancient orders, however, were still abundant, marsupials, insectivores, rabbits and rodents. The hedgehog and mole families were already recognizable amongst the insectivores; herbivores could be distinguished as peccaries and widespread tapirs. Carnivores, too, were becoming modern; dogs, weasels and sabre-tooth tigers roamed over the north. The mammalian fauna was becoming complex. All these new families were widespread at their first appearance in oligocene rocks so that their exact place of origin is not known, but only that they spread widely and rapidly. There must therefore have been a restoration of the continental connexion which had been presumed broken in late eocene days, to permit this large-scale faunal interchange. But not all groups travelled widely; some of both old and new were confined to one region or another. The paleanodonts and the camels were still characteristic of the Nearctic and they had been joined by the pocket gophers. In the Palearctic the first pigs and deer had arrived. There were also civets, parrots, trogons and ancestors of modern giant salamanders. Because of this differentiation between Palearctic and Nearctic in the later years of the oligocene it has been supposed that after inter-communication had been restored in early oligocene days, a new break had occurred. Not all the differences can be accounted for in this way, however, because the non-migrating paleanodonts and camels had already had an opportunity to spread and had not taken it. They appear to have been restricted by limitations other than sea barriers. However this may be, the zoogeographical regions were becoming more distinct. The fauna of both had similar components, marsupials in the trees, insectivores on the ground, rabbits, rodents and ungulates living on a herbivorous diet, dogs, tigers and mustelids to eat them, birds in the air and amphibia and fish in lakes and rivers. The two regions however, differed in their herbivores. Whilst tapirs and peccaries were common to both, there were browsing camels and gnawing gophers in the Nearctic, grazing deer and rooting pigs in the Palearctic. There were paleanodonts in the Nearctic and the Palearctic had an extra family of carnivores, the civets. Of the other vertebrates, there were axolotls in the Nearctic to correspond with the salamanders of the Palearctic. The parrots and carp had still not spread.

15 million years later, in the miocene with the exception of the paleanodonts these same families were flourishing and becoming more modern in appearance. The differences between Nearctic and Palearctic were increasing

because there was probably little interchange until well on into the miocene
when a land connexion was once more established across some part of the

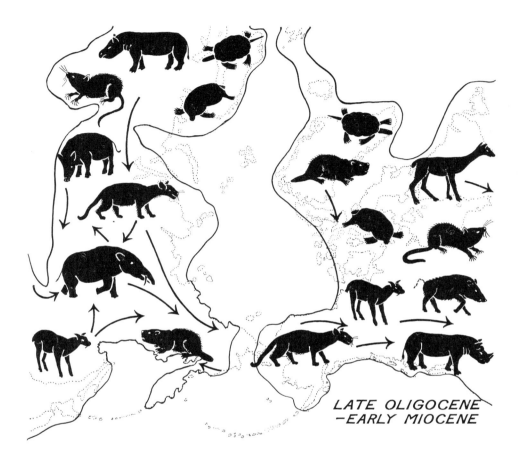

LATE OLIGOCENE
–EARLY MIOCENE

Bering Straits between Alaska and Siberia. Deer, carp and parrots spread
to the Nearctic and squirrels were found throughout the north. New families
appeared in both regions; bovids, giraffes and bears in the Palearctic, pronghorns in the Nearctic. There were Old World monkeys in the Palearctic for
the first time. In many ways the Nearctic and Palearctic had already taken
on their modern characteristics by the miocene. Already axolotls, gophers
and pronghorns were typical of the Nearctic fauna. Holostean fish contracted to their modern, Nearctic, distribution. Pigs were typically Palearctic and only spread to the Nearctic in historic times. In other respects the
differences that existed between the regions in miocene days were differences
that did not last. Neither the camels of the miocene Nearctic, nor the beavers,

giraffes and bovids of the Palearctic were to remain typical of their original regions.

By the pliocene the Palearctic was cut off by the Himalayas from the Oriental region, but it seems to have had a connexion with the Nearctic during part of the period because there was migration between the two. Bears and giant salamanders spread to the Nearctic, camels which had been restricted to the Nearctic since their first appearance in eocene days migrated

MID-PLIOCENE

to the Palearctic for the first time. But the pliocene north also lost some of its earlier fauna. Neither side-neck turtles nor marsupials were any longer to be found there. But in spite of these interchanges and extinctions which increased the uniformity of the two northern regions, many of the old differences persisted and new ones came into being. These differences were due both to extinctions and new arrivals. Thus rhinoceroses which had been widespread, now vanished from the Nearctic; their continuous range was contracting. Hyenas, aardvarks and spalacid mole rats were new to the fauna when they appeared in the Palearctic. None of them ever colonized the New World. And so at the end of the pliocene the two northern zoogeographical regions were still distinctly different. Their differences were not the same as they had been in the previous epoch, nor were they the same as they are today, but they were gradually coming to resemble more closely the modern regions than they had done in earlier days.

The next epoch, the pleistocene, is not long ago in terms of mammalian evolution and most of the modern forms had evolved by then. The most notable events of this time were the ice ages. At the beginning of the Cenozoic it was probably warm and equable over most of the world, but by the pleistocene much of the land of the north was covered by ice. Many northern mammals had already migrated south and by the end of the pleistocene were the only survivors of once widespread groups.

The pleistocene is a particularly interesting period in the records of mammalian evolution. On the one hand many apparently flourishing forms suddenly died out and on the other, giant versions of several well-known mammals made their appearance in various parts of the world. There were giant beavers and mammoths in the north, giant moose in North America and the huge Irish elk in the Palearctic. Throughout the New World there were giant ground sloths; in Australia a giant kangaroo, and in Africa giant pigs. Yet another feature of the pleistocene was the thick fur coat that some of the northern mammals acquired. There were the woolly rhinoceroses of Eurasia and the woolly mammoths of both northern regions.

As the pleistocene ice spread south, many animals unable to survive in the new conditions either died out altogether or took refuge in the southern continents. Mammoths and elephants as well as tapirs and sabre-tooths, most of which had been typical of the northern fauna for many millions of years, were amongst those that suffered from these conditions. Mammoths and sabre-tooths vanished from the world for ever, but the tapirs and the more recently evolved elephants had already established colonies in the south. The southern colonies of the tapirs were at the extreme western and eastern ends of their range, so that their pleistocene extinction in the north left them with their modern pattern of widely spaced discontinuous distribution. Elephants although also members of the Nearctic and Palearctic in the pliocene were only established southwards in the Old World. Other families which suffered the same fate of northern extinction in the pleistocene were already restricted to the Old World. Thus monkeys, rhinoceroses and aardvarks disappeared from the north of their range. Camels on the other hand which were shared by Nearctic and Palearctic disappeared only from the Nearctic. For three Cenozoic epochs they had been typical of the Nearctic, only spreading south and west in late pliocene times. But in the pleistocene they vanished from their region of origin, from the centre of their range and so became discontinuously distributed. In contrast, the differential extinction of the peccaries, confined them in the pleistocene to the New World. In spite of the large-scale extinction in the north, a few families extended their range. Bovids and salamanders spread from Palearctic to Nearctic. Other families retained their place in the northern fauna amongst which were many of the regionally characteristic families. The holosteans, axolotls, gophers and pronghorns remained in the Nearctic; spalacids, hyenas and pigs in the Palearctic. No new families except the Hominidae appeared in the pleistocene. By the end of the last glaciation the faunas of Palearctic and Nearctic had taken on the characteristics that distinguish them today.

Gradually then through the Cenozoic what was a uniform northern fauna separated into Nearctic and Palearctic. As new animals evolved, some like the bears, salamanders and carp spread throughout the north, others like the pronghorns and axolotls remained in the region of their origin, whilst others like the camels and parrots spread from one region to the other and then died out in the region they had come from. Present day distribution depends not only on factors operating today but also on these same factors operating in the past when both the animals of the world and the continental connexions of the world were different.

8

Southern Regions

Intermigration

The Cenozoic evolution, spread and contraction of modern vertebrate families in the north had its effect on the fauna of the southern continents. Assuming that most land vertebrates only rarely cross sea to establish themselves in new places, the newly evolving northern families could only migrate into the southern continents at times when these had land connexions with the north. This led to differences in colonization of the southern continents, for they did not all have northerly connexions at the same time. Furthermore, differences were caused by differences between Palearctic and Nearctic. At times when regional differences between these two were marked, different faunas would be available for southwards migration from them. Thus South America could only receive animals that were already in the Nearctic, Africa only those that were in the Palearctic. Connexion with the north only, but at different times accounts for most of the distributional characteristics of the southern continental faunas.

In contrast to the general theory that land vertebrates require land by which to travel from one continent to another, some animals seem able to migrate along a chain of islands if the sea between them is not too wide. This applies in particular to reptiles, birds, bats, rats and some of the small arboreal mammals. Several instances of this type of colonization are suspected.

Although it is assumed that the majority of mammal families originated in the north, this is known not to be true of all. Several have originated in South America and others in Africa and Australia. Sometimes the southern families spread successfully into the north, when given the opportunity, more often they remained in their southern homes, or became extinct when new northern forms arrived.

Neotropical

At the end of the Mesozoic and for a short period of the paleocene, South America was joined to North America by a thin neck of land in much the same place as the one that joins them today. At this time there were marsupials in the northern hemisphere and a few early placentals. Some of these mammals migrated into the southern continents probably with some of the birds, reptiles, amphibia (pipid toads perhaps) and fish (characins for example) typical of the time. Lungfish, side-neck turtles and crocodiles were already there by the cretaceous.

Shortly after this early invasion, some time in the paleocene, South America was cut off from the north by the submergence of the narrow connecting corridor. South America could no longer receive migrants from the north. Those it had already received evolved in isolation to fill the available habitats.

LATE
CRETACEOUS
– EARLY
PALEOCENE

These earliest of South American mammals were representatives of only three groups from the north, the widespread marsupials and condylarths and the Nearctic paleanodonts. Once isolated in the Neotropical region, the marsupials diverged widely, not only into the tree-living and mouse-like forms of the modern region, but also into a large group of carnivorous animals, ranging in size from that of an opossum to that of a bear. One of the later of these marsupial carnivores was like a sabre-tooth tiger. Marsupials in fact were occupying many of the carnivore places in the fauna in ways parallel to their placental counterparts which were absent from the Neotropical at this time. The condylarths and paleanodonts were the only placentals present in the Neotropical at this time and neither group was carnivorous; the paleanodonts were insectivorous and the condylarths herbivorous. The paleanodonts evolved subsequently into some of the most typical of present-day Neotropical animals, armadillos, sloths and anteaters. But they too were more diverse in some epochs than they are today. For instance in the pleistocene there were giant ground sloths, giant armadillos and glyptodons. The condylarths also differentiated widely, but unlike the other two early immigrant groups, they have left no modern descendants. Some became strikingly similar to other lines of mammals. There were amongst them a 'horse' and good imitations of a rhinoceros and a hippopotamus, as well as many rat-like forms. This extensive group of early placental herbivores only reached this degree of diversity in South America. South America was isolated from the north just at the right time for these forms to diversify widely. Isolation kept out the placental carnivores who would have eaten them, and the northern herbivores who might have ousted them.

It was a little later in time, after these first mammalian groups had become established, that the ancestors of the South American monkeys and rodents arrived. It is possible that they joined the ancient immigrants in the oligocene, and may, therefore, have arrived by island-hopping across the sea channel. Being small and arboreal they would have been more suited to this type of

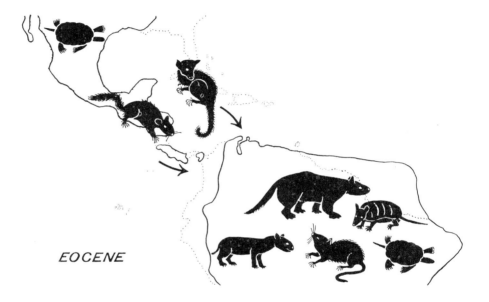

EOCENE

spread than other northern animals of the time. But whatever the date of their first arrival, and by whatever means they effected it, by late Cenozoic days they too had evolved into many different types.

The Neotropical fauna of mid-Cenozoic days was therefore composed of recently arrived placental rodents and arboreal primates, ancient paleanodont insectivores, large herbivores of ancient condylarth stock and a multitude of carnivorous, arboreal and rodent marsupials. Already, therefore before the pliocene, the Neotropical had many of its modern faunistic features, families of marsupials, edentates, monkeys and rodents.

Towards the end of the pliocene the Americas were once more joined by dry land and remained connected until the cutting of the Panama canal in recent times. At first the connexion was probably no more than closely lying islands and the invasion by modern placentals from the north was slow. The raccoons may have been the first to make the journey south. But as the land connexion became continuous there was a speeding up of the southerly flow. Once again, only those animals that had already established themselves in the Nearctic were available for this southerly dispersal. Thus, although many

modern carnivores went south at this time, there were no hyenas amongst
them because hyenas had not spread over from the Palearctic, and there are
therefore no hyenas in the Neotropical today. Cats and bears spread south
and because they were more successful than their marsupial counterparts,
gradually supplanted them. Gradually, too, the special South American
herbivores disappeared, falling prey to these new carnivores. With their
disappearance, there was an empty place for the new northern herbivores to
fill. Peccaries, deer, camelids and horses were available in the Nearctic for
this take over, but there were no rhinoceroses for they had already gone from
the Nearctic. No giraffes were available because they had never lived in the

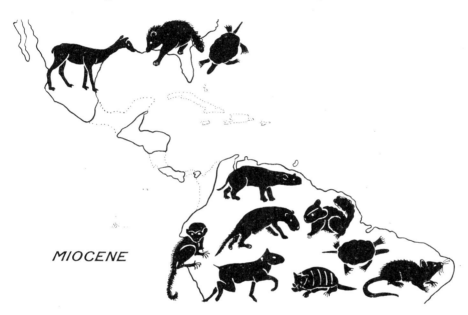

MIOCENE

New World, and neither bovids nor elephants had yet spread across from the
Palearctic to the Nearctic. At the end of the pliocene therefore the Neotropical had acquired a further batch of characteristic animals, amongst them
camels and peccaries.

By the pleistocene, otters, dogs, weasels and sabre-tooths as well as many
of the smaller mammals had come to join the first invaders, and tinamus
were already established. Even as late as this, another characteristic Neotropical animal had arrived, the tapir. But neither beavers, bovids, nor
elephants spread south from the Nearctic although mastodons had by this
time reached northern South America. The pleistocene fauna was soon complete and has persisted to the present day with only one or two exceptions. At
the end of the pleistocene, horses, sabre-tooths, giant edentates, mastodons
and the last survivors of the marsupial carnivores and of the South American
herbivores died out. The elimination of the marsupials in the pliocene and
the camels and tapirs in the pleistocene from the northerly end of their
ranges, left colonies of these animals isolated in the Neotropical.

Although most of the migratory movements had been southwards during the Cenozoic, a few animals made their way north during the pliocene and pleistocene when there was a corridor of dry land to use. Of these, giant

LATE PLIOCENE

armadillos, glyptodons and giant ground sloths had only a short evolutionary life in North America, but the later arrivals in the north, the opossum and the nine-banded armadillo, colonized the Nearctic successfully and survived. So, too, humming birds and tanagers went north from their Neotropical home, but the majority of the Neotropical animals which had evolved during the period of Cenozoic isolation, stayed where they were.

Thus the modern fauna of the Neotropical region can be accounted for by its early isolation when it contained only a few mammals, represented today by marsupials and edentates; the diversification of this fauna, together with rodent and primate immigrants, during the Cenozoic; and the acquisition of a number of modern mammals from the Nearctic during the pliocene, some of which subsequently became extinct at their northern source.

Ethiopian

Fossils of cretaceous and early Cenozoic strata of Africa are rare, and for this reason it has proved difficult to reconstruct the story of the Ethiopian region. Much has depended on guesswork and deductions from what is known of the present-day distribution of families that inhabit Africa. What follows is one possible scheme of the sequence of events which led to the establishment of the modern zoogeographical region. Much may need to be changed when further fossil evidence becomes available.

Cretaceous Africa resembled cretaceous South America and the northern hemisphere in many ways. There were lungfish, crocodiles and side-neck turtles and probably also pipid toads. But beyond this the fossil record reveals little. Some early placental mammals may have been living in the Ethiopian region at this time; insectivores and condylarths were likely inhabitants.

At the end of the cretaceous, Africa probably lost land connexion with the north and became an isolated continent surrounded by sea. The basic fauna could evolve in isolation just as it would do a little later in the Neotropical, but the basic Ethiopian mammal fauna differed from that of the Neotropical. Ethiopian insectivores and condylarths contrasted with Neotropical marsupials, paleanodonts and condylarths.

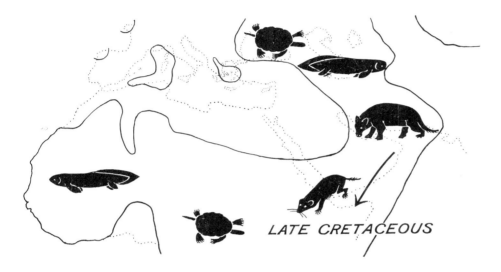

LATE CRETACEOUS

During the period of isolation, which lasted at most until the late oligocene, lemur-like members of the primate stock may have crossed the sea from the Palearctic. Lemurs being small and arboreal and the distance short, this is not improbable. It has already been suggested as an explanation for the arrival of primates in South America. It will later be put forward as a reason for the presence of lemurs in Madagascar.

During the early epochs of the Cenozoic, then, there were insectivores, condylarths and primates in Africa. The sequence of insectivore evolution is unknown, but at some time the modern African families must have diversified. The golden moles, the elephant shrews and the otter shrews may have originated in the continent in which they are found today. By the oligocene the condylarths had differentiated into several sorts of large herbivore as they had been doing in South America at the same time. But the results were different. Amongst the Ethiopian herbivore stock there were already elephant and mastodon ancestors, and a relatively huge animal, *Arsinoitherium*, with paired horns on the end of its nose. Old World monkeys were becoming distinct, differentiating from their lemur-like ancestors. Thus it seems that they preceded in time the monkeys of the Neotropical, although they may both have been derived from similar northern ancestors.

By the end of the oligocene Africa had gained a connexion with the
Eurasian continent by way of a tract of forested land through what is now
Arabia. There may have been an additional connexion further west. Late

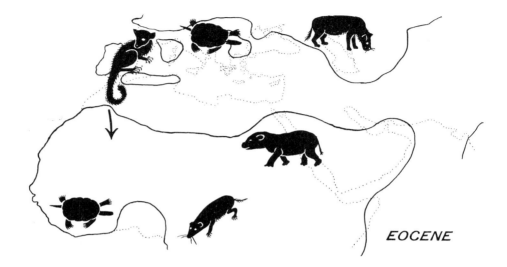

EOCENE

oligocene and miocene placentals from the Palearctic could invade Africa.
Modern carnivores and rodents poured south. Cricetid mice, cats and dogs
and the aardvarks occupied Africa at this time. Later on, towards the end of
the miocene and through the pliocene, the corridor between Africa and Asia
lost its forests and became dry. Whilst the corridor was changing in this way
there was an influx of antelopes and many other animals that are typical of the
plains of modern Africa. Palearctic giraffes, rhinoceroses, hyenas and murid
mice made their way to Africa. Bears did not evolve in time to get into Africa
before the corridor became too dry for them, but it is not easy to see why the
deer failed to make the journey. Presumably they had a too northerly range
during the crucial period before the land between Africa and Asia became
unpleasantly arid.

Whilst the southerly invasion was going on, some of Africa's own products
spread northwards. Mastodons and elephants left Africa and spread first into
the Palearctic and then made their way into the Nearctic and Oriental regions.
Much later, in the pleistocene, man-like apes may have gone north from
Africa and spread round the world. In spite of their African origin, modern
elephants had to recolonize Africa from the north. The early African forms
died out. And the same story may apply to modern man.

Many of the original inhabitants of Africa must have found it hard to
survive when the new mammals came down from the north after the mid-
Cenozoic corridor had been made, and it is probable that many were considerably reduced in number. Some of the insectivores and primates may
have suffered this fate.

As the Asia-African corridor became more arid it formed an effective barrier for almost all further exchange of fauna. By the pleistocene, Africa was isolated again from the rest of the world and her fauna evolved into the typically African forms of today.

The Ethiopian fauna can be accounted for then by colonization from the Palearctic during the cretaceous and the late oligocene to pliocene, with differentiation of the native fauna during the periods of isolation. The fauna is composed of insectivores, the results of Cenozoic isolation, and of a large number of northern families of which some are widespread but others remnants of families that were left in the south when the ice sheets descended. It is in these southern remnants that the Ethiopian resembles so markedly the Oriental region; the rhinoceroses, monkeys and elephants for example. Although these families have today an identical distribution, restricted to the Old World tropical regions, their distribution histories varied. Elephants, sprung from African stock, went north to the Palearctic and Nearctic and then south again to the Ethiopian and Oriental. Their extinction in the north in the pleistocene left them with their Ethiopian and Oriental colonies. Surprisingly their southern migration never took them to the Neotropical.

MIOCENE

Monkeys, following the same route, never even got to the Nearctic. Their extinction in the Palearctic left them too with only Ethiopian and Oriental populations. Rhinoceroses were originally northern animals, but as they spread south in the Old World, their range contracted away from the Nearctic, and subsequent Palearctic extinction led to the same result once more.

There are similarities between the Ethiopian region and the Neotropical, both in climate and some of the verterbrate fauna, but the major part of their faunas is different. Many of the differences can be accounted for by the fact that in the cretaceous the two regions had received different components of the northern fauna, by the fact that they later regained connexions with the north in different epochs, and by the fact that when they did regain these connexions the connexions were with different northern regions both of which had already taken on some of their modern characteristics.

The fauna of the Ethiopian region differs from that of its Palearctic neighbour both in climate and because it was not in continuous connexion with it through the Cenozoic.

Oriental

The Oriental region has a less complex history than most of the other regions. Once it had become part of the Asiatic mainland towards the end of Mesozoic times when the Tethys Sea dwindled away, it shared its fauna with the north.

Well on into Cenozoic days there was no differentiation of an Oriental fauna. Unfortunately the story of the Oriental region suffers like that of the Ethiopian from lack of fossils during the early epochs of the Cenozoic. Although it can perhaps be presumed that there were a number of frog and toad families, side-neck turtles, crocodiles, marsupials and insectivores in the Oriental cretaceous, as there were in the Palearctic at the time, there is no confirmation of this in the rocks. As a knowledge of the range of these families during the cretaceous is crucial for an interpretation of the colonization of Australia this is doubly unfortunate.

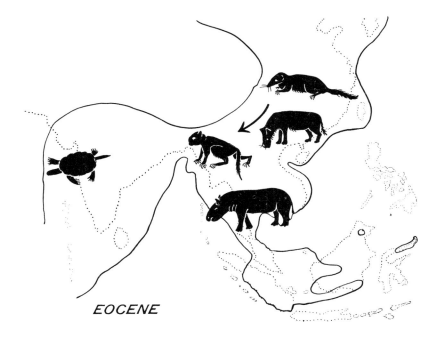

EOCENE

Assuming that the Oriental did indeed share its fauna with the north until the end of the oligocene, it would have had primates, lagomorphs, rodents, tapirs, rhinoceroses, deer and pigs. Of the northern rodents the beavers, however, never spread south as far as the Oriental region.

Ancestors of the Oriental tarsier lived in the eocene north, and a little later, oligocene, a primitive tree shrew inhabited the Palearctic. This is all that is known of two of the most characteristic modern Oriental families. They must have spread to the Orient at some time during the first part of the Cenozoic, eventually dying out in the north. Spread from the north, followed by extinction in the north is the explanation of most of the Oriental fauna.

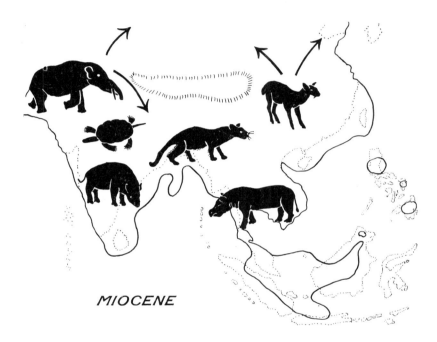

MIOCENE

In miocene days great earth movements had initiated a period of mountain building which was to last well on into the pliocene. During this upheaval the Himalayas were built up. At first they lay as a comparatively small range between the Palearctic and the Oriental. At this time they cannot have hindered the free interchange of animals between the north and south. Elephants which had only recently spread north from Africa found their way to the Oriental in the miocene, and it is possible that the Old World monkeys followed the same course. Gavial crocodiles which had been inhabitants of the Palearctic spread southwards to both Africa and India, but for unknown reasons they did not maintain themselves in Africa. Giraffes too spread from the north into both of the Old World southern regions, but they did not survive in the Oriental.

By the end of the pliocene two events had occurred, which were the most significant of all for the modern Oriental fauna. The building of the Hima-

layas was complete, providing a barrier between Palearctic and Oriental, and
the climate of the north was cooling.

PLIOCENE

As a result of these two physical changes, the Oriental fauna was isolated
and could differentiate from the Palearctic, and many of the families formerly
with an extensive range became extinct in the north. Thus the Oriental, like
the Neotropical and Ethiopian, came to contain the surviving tropical
elements of once widely distributed animals. The disappearance from the
north of the tapir family left colonies in the Oriental and Neotropical, but in
most other respects the Oriental retained mainly Old World elements and
in this resembled the Ethiopian, with rhinoceroses, monkeys and elephants
for instance. This extinction of families in the pleistocene north and their
continued existence in the south of their range provides the main cause for
the difference between Palearctic and Oriental regions, whilst at the same
time providing the reason for the similarities between Ethiopian and Oriental.
Oriental and Ethiopian represent the tropical fauna of the Old World.

Australian

Australia differs from the other southern continents in having had no connexion with the north during the whole of the Cenozoic. As no early Cenozoic
fossils are known from Australia it is difficult to tell when she acquired the
ancestors of her modern fauna. Guesswork suggests that most of the land
vertebrates were occasional immigrants at various times through the

Cenozoic from the Asiatic mainland along the islands of the Malay Archipelago.

Already by the cretaceous there were lungfish and crocodiles in the Australian region and probably also liopelmid and other frogs including hylids; but, curiously, no early remains have been found of the side-neck turtles in Australia, although they were otherwise widespread at the beginning of the Cenozoic. Ancestors of the egg-laying mammals may have been in Australia by the cretaceous although there is no evidence of this. Their remote ancestors are known from Palearctic triassic and jurassic days.

It is not known, either, when the marsupials arrived in Australia, but it is thought to have been in early Cenozoic or cretaceous days. How they got to Australia and why they, rather than their placental contemporaries, achieved this distinction is equally unknown. Earlier theories assumed both that

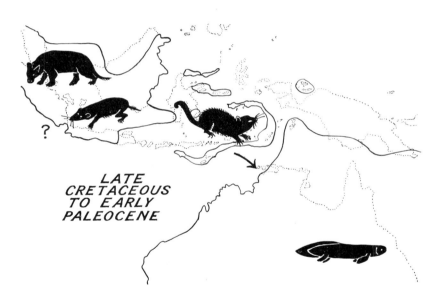

LATE
CRETACEOUS
TO EARLY
PALEOCENE

marsupials preceded placentals in time and that Australia had a land connexion with Asia in the early Cenozoic. Thus marsupials colonized Australia before the evolution of placentals. The land connexion with Asia broke down and the marsupials were isolated from later influxes of placentals. It is unlikely that this can be true. There is no evidence that marsupials were in existence before placentals and there is no evidence that Australia has had any land connexion with Asia since before the cretaceous.

How can the mammal fauna of Australia be explained? It could be supposed that although marsupials did not precede placentals, they increased in numbers and range at an earlier date. They might then have spread through southern Asia and along the island chain to Australia before the placentals had spread so far south. Once the marsupials were established in Australia it must be supposed that the stretches of sea between the islands increased in width and prevented the spread of the placentals along the islands. There is some evidence that this latter supposition is true but there is no evidence to support the first part of the proposition, that marsupials were more widespread than placentals. Neither marsupial nor insectivore fossils are known from the Oriental at this time.

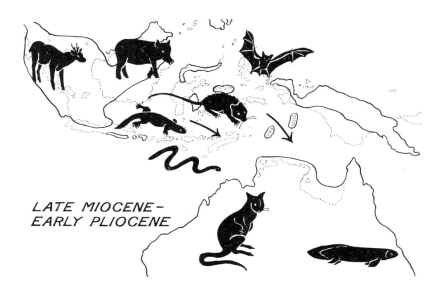

LATE MIOCENE-
EARLY PLIOCENE

Simpson (1940) has suggested an alternative. He has supposed that there were both marsupials and insectivores in southern Asia in the cretaceous and that a string of islands lay between the Oriental mainland and Australia. He further supposes that because the marsupials were arboreal animals they would be more likely to make the journeys from island to island than the ground-living insectivores. Later the journeys became more difficult, and there is geological evidence to support this, so placental migrants arrived only on the rarest occasions. Established marsupials would have been able to prevent successful colonization by rare placental visitors.

However they got there or wherever they came from, the marsupials found the Australian region a suitable home. They were able to evolve into all sorts of different forms and parallel many of the placental types which occur in other parts of the world. Thus, today, Australia has marsupial carnivores, herbivores, rodents, insectivores and flying arboreal mammals.

The Australian region is not entirely free of native land placentals. There are murid rats and there are bats. The Australian murids probably arrived from Asia in the miocene. Pigs and dingo dogs were later arrivals.

Although a few placental mammals got into Australia, no marsupials spread back to Asia. The only westerly spread has been into a few islands of the Malay Archipelago as far as Celebes.

It is likely that most of the other native land vertebrates of Australia were also casual migrants. The occasional frogs, though rare, replaced the earliest liopelmids. An occasional turtle, agamid lizard and snake travelled from Asia from time to time. Birds arrived at different times, mainly from Asia.

The Cenozoic history of Australia differs greatly from that of other southern regions. Unlike any other region it has been isolated for the whole period. Australia and the Oriental regions are at opposite extremes: the one always isolated through the Cenozoic, the other not isolated until the end of the pliocene. In between are the Neotropical and Ethiopian regions with a comparatively long period of Cenozoic isolation and with northern connexions both before and after this isolation.

This interpretation of the arrival of the modern faunas of the six zoo-geographical regions of the world is in accordance with the theory of the permanence of the continents, a theory that permits the making and breaking of only narrow necks of land between the six continents, in places where the sea is shallow today or where land connexions still exist. This theory has been shown to account satisfactorily for most of the Cenozoic patterns of distribution of land mammals, but some problems remain unexplained and little has been said of other land vertebrates.

9

Land Bridges

Gondwanaland

Although the theory of the permanence of the continents has provided a satisfactory basis for the interpretation of most Cenozoic distribution, there are other ways of explaining it. There are other ways of accounting in particular for the presence of marsupials in Australia and the New World and for the presence of lemur-like animals (true lemurs and lorises) in the Ethiopian region, Madagascar and the Oriental region, and for the even more difficult distributional irregularities shown by earlier vertebrates, fish, amphibia and reptiles.

The other ways are provided by the theories of land bridges and continental drift.

Ever since the early part of the last century, difficult problems of animal distribution have been explained by an appeal to land bridges. These land bridges are not the same as those narrow connexions of land demanded by the theory of the permanence of the continents and restricted to regions where the

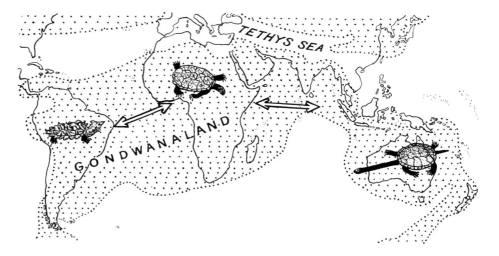

sea is shallow. The theory of land bridges demands bridges of continental size, stretching across what are now deep oceans, serving as a means of communication from one present-day continent to another. Some time during the Cenozoic, when they had served their purpose, these land bridges were supposed to have subsided beneath the oceans leaving no trace of their former existence.

For a long time it was thought that at the beginning of the Cenozoic and earlier, the land of the world was divided into two halves, a northern land mass taking in all of Eurasia, Greenland and North America, and a southern land mass, called Gondwanaland which included the three southern continents, New Zealand and Antarctica. The two land masses were separated by an extensive Tethys Sea. The continuity of the southern land was thought to account for the distribution of lungfish, side-neck turtles, flightless birds and even marsupials, all those animals in fact which are similar in the three southern continents today but which are not widespread through the rest of the world. However, the southern regions do not resemble one another enough to substantiate this theory. The Neotropical and Ethiopian regions are more like one another than either is like the Australian region. Modifications were made, therefore, in the continental bridge theory. Gondwanaland was abandoned, at any rate as an explanation of post-cretaceous distribution, and individual connexions were postulated. South America and Africa were joined by a huge South Atlantic land bridge leaving the other southern continents free, or alternatively, Africa, Madagascar and India were joined into the continent of Lemuria. South America and Australia were joined through a habitable continent of Antarctica. There were many other land bridges to account for local difficulties of animal and plant distribution, but these were the largest and the most widely used.

South Atlantic Bridge

The South Atlantic continent at its greatest was supposed to have included besides Africa and South America, the South Atlantic islands of St. Helena, Tristan da Cunha and Ascension. Across this continent were supposed to have spread the lungfish, characin fish, pipid toads, side-neck turtles, ostriches, porcupines and monkeys; all those animals that Africa and South America have in common today but which are absent from most of the other regions. However, the monkeys and porcupines of the Old World and the New World are no longer thought to be closely related to one another, and it is likely that this is also true of the ostrich-like birds. Although not closely related, they are thought to have evolved to look like one another because they live under the same sort of conditions, just as many marsupials look like their placental counterparts. But even if they are related closely, belonging to the same families, to account for the distribution of mammals and birds by the South Atlantic land bridge would mean keeping the bridge until oligocene days (when monkeys and porcupines first appear in the rocks). By this time edentates and South American ungulates were well established and yet they do not seem ever to have been in Africa. A late land bridge across the South Atlantic raises as many awkward questions as it solves and must almost certainly be rejected. It is still possible however that a much narrower connexion in late Mesozoic days might have been responsible for the spread between the two continents of the turtles, toads and freshwater fish which they have in common. But fossils of side-neck turtles, lungfish and characin fish are known from cretaceous strata of the north and they could therefore have reached the southern continents by a southerly spread, making a direct east-west connexion between Africa and South America unnecessary. Moreover, the other land vertebrates of the late Mesozoic, the amphibia and reptiles, of Africa and South America are not more like one another than they are like those of North America of the same period. Striking similarities between the fossils of Africa and South America are not found until as far back as triassic and permian days and at this point in geological time it becomes difficult to deduce anything about the extent of the continents because of the rarity of fossils.

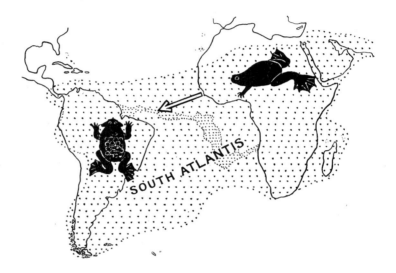

The zoological evidence then, suggests that the continental land bridge of the South Atlantic was unlikely to have existed later than Mesozoic days and that it is not essential for explaining any of the faunistic similarities between present-day Africa and South America. But the final decision for or against such a bridge should be based on geological evidence. Was there land in the South Atlantic or was it always ocean?

The ancient rocks of eastern South America and west Africa show resemblances to one another, though not more than they do to similar rocks in other parts of the world. Further, the South Atlantic is an unstable area, likely to throw up ridges and islands as a result of volcanic activity, which might seem to support a bridge theory, though it does not even suggest the position or the extent of such a bridge. In contrast, some geologists claim that there is evidence that the floor of the Atlantic has been an ocean floor since jurassic days. This would put the latest possible date for an extensive bridge back to the jurassic. This is as far as geological evidence goes. It does not provide a decisive answer. The decision after all has to be made on biological evidence and biology has an alternative explanation of northerly origin and southerly spread along narrow connexions to the permanent continents of

Africa and South America, which is at least as good, and some think better, than an Atlantic bridge. It does not preclude a narrow jurassic bridge nor other bridges in earlier epochs between Africa and South America. Judgment on these must be reserved for further evidence to become available, but the evidence already available makes a Cenozoic bridge of continental dimensions unlikely.

Lemuria

The second big land bridge was the supposed continent of Lemuria. This was said to lie between Africa, the island of Madagascar and India. It was called Lemuria because it was invented in the first place to explain the curious distribution of 'lemurs'. The 'lemurs' were imagined as originating on Lemuria at the end of the Mesozoic and spreading to Africa and the Oriental region, where they are represented today by the lorises. The original lemurs were left on Madagascar when the rest of Lemuria sank below the seas sometime in the Cenozoic.

The reclassification of the primates in recent years has made this interpretation untenable. It is now supposed that although both lorises and lemurs derived from the early primate stock, which was widespread in the north in paleocene and eocene days, they are not otherwise very closely related to one another. Thus it is supposed that the ancestors of the modern lemurs, whose fossils are known in the north, spread south to Madagascar early in the Cenozoic. Only very much later did a second wave of similar primates spread

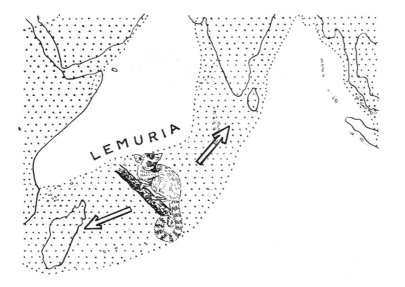

south as lorises to colonize the Ethiopian and Oriental regions becoming extinct in the north in much the same way as the rhinoceroses and monkeys. This would explain the close relationship between African and Asiatic lorises and their remote relationship with Madagascar lemurs much better than could a continent of Lemuria. Lemuria only complicates the zoogeographical problem as far as 'lemurs' are concerned for it implies that the most distantly related should be the two loris families, at opposite ends of the Lemurian continent. This is obviously unsatisfactory.

Other affinities that Madagascar has with the Oriental region can also be accounted for without Lemuria, and there is no conclusive geological evidence to decide the issue for or against the theory. Such as it is, the geological evidence suggests that Madagascar has been isolated from all continents during the whole of the Cenozoic and continental Lemuria should be rejected.

Antarctica

The third of the big land bridges was supposed to have passed from South America through the Antarctic to Australia. It was supposed to account for the presence in South America and Australia of marsupials, hylid tree frogs,

flightless birds, side-neck turtles and lungfish. It is now almost certain that the flightless birds are not closely related to one another and are classified in separate orders, so their problems of distribution are different and need not be discussed in this context. Lungfish and side-neck turtles were widespread in the cretaceous and their discontinuous distribution today is therefore better accounted for by extinction in the north of formerly world-wide groups. There remain the marsupials and the amphibia of the two regions.

Neither marsupial fossils nor living marsupials or hylid tree frogs occur in the Oriental region, although marsupial fossils are known from many other parts of the world. How did marsupials and hylid frogs get into Australia? Is it bad luck that there are no fossils of these groups in the Oriental region, or did they get to Australia direct from South America? But the marsupials were contemporaneous with edentates and South American herbivores and yet there is no evidence for the occurrence of either of these orders in

Australia. They may simply not have spread, but even so the mammal populations of the two regions have more differences than similarities, a result which would hardly be expected if the regions had been connected by a bridge in the early Cenozoic. The same argument applies to the birds, reptiles, amphibia and fish. The likenesses are rare, side-necks and fish which are not even in the same family in the two regions, and hylids which are; the differences are frequent, Asiatic agamid lizards in Australia, iguanids in America, the only teleost freshwater fish family of Australia confined to the Old World. An Antarctic land bridge, like the other continental bridges, raises as many problems as it solves if applied to the distribution of Cenozoic and late Mesozoic animals only. It is possible, however, that the Antarctic was once less cold than it is today, and might have supplied at least stepping-stones from South America to Australia for some early animals, perhaps for invertebrates which show considerable resemblances between the two continents. Fossil beech leaves have been found in the Antarctic, but so far no animal fossils. Further paleontological work may throw light on the problem, but once again crucial evidence must come from non-biological sources, and once again it is not yet available.

Failing definite geological evidence one way or another the question of land bridges has to be decided on biological evidence, and yet it was for biological reasons that the bridges were first invented. This makes the decision difficult, but it is probably fair to conclude that as far as late Mesozoic and Cenozoic animals are concerned, the continental land bridge theory is unnecessary. What happened before this is an open question.

Cenozoic land bridges of continental size are not fashionable today, but in the nineteenth century they were considered essential to explain every small difficulty of animal and plant distribution. A bridge was invented to connect Portugal with Ireland, another from South America to the Galapagos Islands, and others from the continents to every island in the world. Today only those narrow corridors of land for which there is geological evidence, or which do not stretch the imagination too far, such as the Panama isthmus, Arabia and the Bering Straits bridge, and on a smaller scale the channel bridge between England and France, are generally believed in. The Cenozoic distribution of animals is usually explained today without vast continental land bridges.

10

Continental Drift

At the beginning of this century Wegener (see du Toit 1937) put forward the theory of continental drift to account for the stranger facts of animal distribution. It did not require new land areas from under the sea like the land bridge theory. The total land mass remained the same as today, but the continents themselves changed their positions.

Wegener believed that the continents were made up of a comparatively light material, called sial because silica and aluminium formed a large proportion of it, and that the floors of the oceans were made up of the much

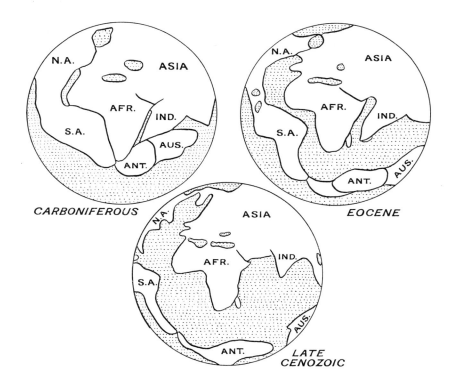

heavier sima, silica and magnesium. He believed that the light continents would float on the lower and heavier earth crust like icebergs. With this hypothesis in mind, and from a close study of the map of the world, he decided that it was possible that all the continents had once been joined together in one block and had then drifted apart to their present positions. This block of land would have existed in the carboniferous, but soon after would have started to break up and the continents would have floated away from one another. By the eocene the Atlantic Ocean was forming, and South America was losing touch with Africa. Australia and South America were still joined through the Antarctic, and not until comparatively recent times did Australia take up her present position. This was the theory of continental drift, and it had been put forward mainly to account for the biological similarities of the southern continents.

Wegener supposed that in the early days when all the continents were joined together they would have shared the same fauna and this was taken to account for the presence of flightless birds, side-neck turtles and lungfish in all three southern continents. A little later, once Australia and the Antarctic had become separated from the African part of the block, except by the long way round of South America, the fauna of the southern continents would have become differentiated into two main components. Australia and South America would share one of these, marsupials and hylid tree frogs being included, and South America and Africa would share the other, ostriches, pipid toads and characin fish for instance.

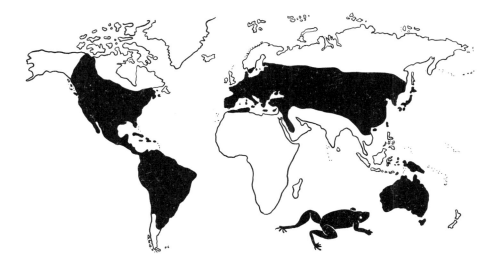

It is an attractive idea. South America and Africa would fit together very neatly, and there is no reason to reject the Antarctic connexion out of hand. But evidence for the drift theory is conflicting. The Atlantic and Indian Oceans may date from jurassic days at least, and this would make the drift theory necessarily a much earlier happening than Wegener estimated. Doubts have been cast on the sial-sima theory, since parts of the floors of the ocean are now known to be sial like the continents. On the other hand recent work on rock magnetism has brought up the whole problem of continental drift in a new light and further information must be awaited (see Irving and Green 1957).

If this modern work proves the drift theory to be correct, continental drift might be the answer to some of the difficult problems of discontinuous distribution. It might account for the distribution of some of the frogs and fish. But in its turn it will raise a host of new problems. Why if the continents were so close together at the end of the Mesozoic do the southern lands not show more similarities than they do ? Although some of the South American fish, frogs, reptiles and birds are very much the same as the African ones, most of them are remarkably different. There are no mormyrid fish, poly-pedatid tree frogs, chameleons or agamid lizards in South America. There are no hylid tree frogs, electric eels or iguanid lizards in Africa. So, too, although Australia resembles South America in some ways, there are also some oustanding differences. Australia has no edentates or condylarth descendants, and yet their ancestors might have been expected to migrate as the marsupials are supposed to have done on the drift hypothesis, from South America to Australia in the paleocene. Australia has no South American lizards, but instead agamids and varanids which are typically Asiatic. The Australian vertebrate fauna is hardly consistent with a continental connexion of any kind. It is typical of a fauna acquired by chance island-hopping. An explanation of these problems will have to be found if Wegener's drift theory proves correct. It is likely that even if the continents were in a close compact mass and drifted apart, this happened considerably earlier than Wegener supposed. It could therefore explain the invertebrate distribution and even the freshwater fish, but the more conservative arguments involved in the theory of the permanence of the continents would still be needed to explain the dispersal of the mammals.

The argument is circular once again. A hypothesis is formulated (drift) based mainly on the facts of animal distribution and then the facts of animal distribution are fitted into the hypothesis, or cannot be fitted as the case may be, and all the time the physical evidence is not consulted. And yet physical evidence is necessary before a decision can be reached. So far, however, the modern work on rock magnetism has not been able to show whether the continents were actually closer to one another in an east-west direction than they are today. The general opinion seems to be that if the physical inter-pretation is correct, there has been rather a northward drift of Africa, India and Australia. This would make the Antarctic land bridge so much the more probable, but sheds little light on the rest of Wegener's hypothesis except to make drifting continents a possibility.

But it seems easier to make animals move round the world and fill perman-ent continents than to make the continents move round to collect them.

The theory of the permanence of the continents has been preferred as an explanation of late Mesozoic and Cenozoic vertebrate distribution because it is the simplest of the theories and because it raises fewer difficulties than the others. What happened in the early Mesozoic and the Paleozoic is much more difficult to guess, and at present there is so little evidence on which to base a theory that the question must be left open. It is hoped that increasing knowledge of the distribution of Mesozoic land reptiles may provide a basis for a theory of land distribution in triassic and permian days.

If geologists find that either drift or land bridges can be proved on physical evidence, then the zoological facts will have to be looked at again. But while the biological facts are the main evidence there is no conclusive reason why the theory of the permanence of the continents should be rejected.

Part Four

ISLANDS

11

Island Patterns

Wallace's Line

When the world was first divided into six zoogeographical regions, in the middle of the last century, it was found difficult to know where to draw the line between the Oriental and Australian regions. The main land masses of the two regions are separated from one another by several strings of islands, the Malay Archipelago. Which islands of this big group belong to which region?

It was obvious from the first that the large islands of Sumatra, Java and Borneo belonged to the Oriental region. They are separated from Malaya and from one another by only shallow sea, and their animals are very similar to those of Malaya. In the east, New Guinea and the Aru islands are close to Australia, within another shallow sea area, and have many animals in common with that continent, including some marsupials and cassowaries. They can therefore be assigned to the Australian region. This leaves the Philippines, Celebes, the Moluccas, Timor and several other groups of smaller islands to be fitted into the zoogeographical classification.

In the middle of the nineteenth century an English naturalist A. R. Wallace was working amongst the islands of the Malay Archipelago, collecting mammals and birds, insects and snails, to send home to England. At first he spent all his time in Malaya and Borneo, but one day in the summer of 1856, he travelled from the small island of Bali to the next island, twenty miles away, to Lombok. He was astonished at the difference in the fauna of the two islands. He had left behind on Bali green woodpeckers and barbets and had arrived on an island where the outstanding birds were white cockatoos and honeysuckers. In crossing twenty miles of sea he seemed to have sailed out of the Oriental region into the Australian. This passage from one region to the other had happened much further west than he had expected.

After this experience he visited Celebes, the Kei islands and the Aru islands, Timor, the Moluccas and New Guinea, and he studied the fauna of each new island in the light of his experience on the Bali-Lombok voyage. Island by island they were classified as being Oriental or Australian. When a number of these islands had been classified in this way he was able to decide where the Oriental region ended and where the Australian region began. In 1863 he drew on the map a line which he considered marked out the boundary between the two regions. This line became known, and still is known, as Wallace's Line, after its originator. The line runs between the Philippines and the Moluccas in the north, then south-west between Borneo and Celebes and finally south between the small islands of Bali and Lombok.

Some years after this another line was drawn between the Oriental and Australian regions, because it was thought to divide the two faunas better than Wallace's Line. This new line was called Weber's Line and was based mainly on observations of the mollusc and mammal faunas of the area. Weber's Line runs between the Moluccas and Celebes, and between the Kei Islands and Timor. It is further east than Wallace's Line.

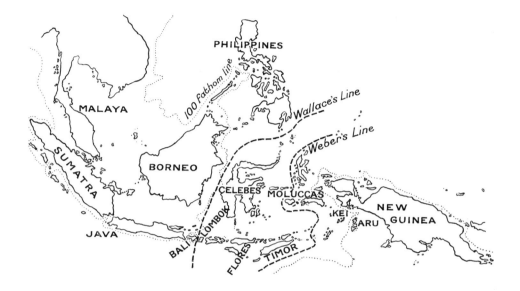

Some people prefer Wallace's Line as the boundary, others prefer Weber's Line. Why is it so difficult to draw a line?

Geologically, Wallace's Line marks off the easterly limit of what was once a land mass joined to Malaya, and Weber's Line more or less marks off the westerly limit of what was an Australian continent at one time. Thus there would seem to be geological evidence for both lines. Islands that once formed part of the mainland could be expected to have a considerable part of the fauna of that mainland. Thus the fauna of Borneo is mainly Oriental, that of New Guinea, Australian. But what is to be done with the islands that lie between the two lines, especially Celebes, Flores and Lombok? These middle islands of the Archipelago are not geologically part of the Oriental region nor part of the Australian region. Many of them were probably under the sea for most of the Cenozoic, thus losing all their earlier plant and animal inhabitants. When they re-emerged, towards the end of the Cenozoic, they were independent islands, never connected by land to either of the continents.

The middle islands have been colonized in a haphazard way by migrations across the sea from both the east and the west. Consequently they each have individual faunas, different from one another and different from either Malaya or Australia. Celebes for instance has very few mammals, hardly any amphibia, no freshwater fish and some peculiar birds. Most of the mammals come from Asia; the bats, shrews, tarsier, macaque monkey, squirrels, murid rats and mice, porcupine, civets, pigs and deer. Others, although belonging to Oriental families have differentiated into forms peculiar to Celebes. There is the dwarf buffalo, the babirussa pig and the black tailless baboon. In the pleistocene there was a small Celebesian elephant. Only the phalangers of Celebes are typically Australian. Presumably all the Celebesian mammals crossed the sea from one or other of the islands of the Archipelago with the result that the fauna is mixed, poor, and not exactly like that of any other island. The other middle islands show similar characteristics, although some have a larger proportion of Australian animals. They do not fall easily into either of the great regions on either side. Some zoogeographers have despaired of ever drawing any one line between the Orient and Australia which would satisfy all biologists and have suggested keeping the two lines, Wallace's and Weber's, and regarding all the middle islands as a separate region, Wallacea. Others insist that one line should separate the regions, and wish to draw it where the fauna is exactly half Oriental and half Australian. The difficulty of doing this is considerable because all groups of animals do not reach the 50:50 mark at the same place. But if one line were to be drawn on this basis it would certainly put Celebes, Lombok and Flores into the Oriental region.

Whichever view be accepted, the important thing to realize is that Wallacea is a transition area between two regions and that it is not unique in this. There

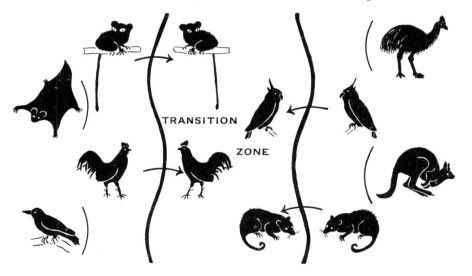

TRANSITION ZONE

are transition areas in Central America where the faunas of the Nearctic and Neotropical overlap, and in north Africa where the Ethiopian and Palearctic overlap. The drawing of a precise boundary line across a transition area between two regions is impossible. Where faunas overlap the area of overlap may be large or small; not all groups of animals may extend to the boundary; the fractions of the main regional faunas involved may be large or small; there may be more overlapping in one direction than another. Therefore all transition areas may vary greatly in extent and composition. Wallacea differs from these two other transition areas however in being an area of sea and islands and therefore its problems are even more complex. Wallacea is not simply a matter of Australian-Oriental crossmigration but of island crossmigration within the transition area itself, further complicated by the independent biological evolution of each island. Wallacea introduces some of the difficulties involved in the classification of islands.

Classification of Islands

There are other islands besides those of the Malay Archipelago that are difficult to assign to a particular zoogeographical region. Islands are of two sorts: continental islands and oceanic islands. It is usually comparatively easy to put continental islands into their zoogeographical region, but oceanic islands are often more difficult.

Continental islands are those which at some time have been part of a large land mass, but which have lost their connexion with the parent land. Through the sinking of a neck of land or a rise in sea level, they have become separated from the continent by a stretch of sea. The sea may be narrow like the Straits of Dover which separate the continental island of Great Britain from the continent of Eurasia, or they may be 100 miles wide like the Formosa Strait which runs between Formosa and the east coast of China.

Continental islands differ from one another in the length of time for which they have been independent islands. Great Britain is a young continental island; Borneo, Formosa and Japan are older.

Most of the land fauna of a continental island can be expected to have reached the island across dry land when the island was still connected to the continent. The length of time for which a continental island has been independent determines the amount of difference between its fauna and the fauna of the mainland. If it is an old island it may be lacking in animals which are comparative newcomers to the mainland, and its own animals may have had time to evolve in different ways from their continental relatives. A young continental island is likely to differ little from the mainland, for it takes many thousands of years for animal species to change to fit a new environment. Continental islands, therefore, are likely to resemble their continents in a general way, and to have features peculiar to themselves only if they have been cut off from the continent for a long time.

In contrast, oceanic islands are islands which have never had a land connexion with a continent. They are of volcanic origin and often many hundreds of miles from the nearest land. The Azores, Bermuda, the Galapagos Islands, Sandwich Islands, Polynesian Islands, St Helena and

Tristan da Cunha are well-known oceanic islands.

The fauna of an oceanic island must be derived from across the sea. The direction from which the fauna comes will be determined to some extent by the prevailing winds and ocean currents, so that in many cases one zoo-geographical region will have been the source of the fauna. But this is not always the case and the fauna may come from more than one region as that of Celebes has done. Sometimes, even, it is impossible to decide on the source of the fauna. In all cases, however, the fauna of an oceanic island is likely to be poor in basic groups, because few will have survived the sea-crossing. Those that have reached an oceanic island, however, may have evolved in ways different from those of their relatives left behind on some continent, and the fauna of an oceanic island is therefore likely to be very individual if the island is of some age.

An oceanic island is likely to differ from the nearest continent in climate, vegetation and fauna. All these environmental differences will lead to the divergence of its inhabitants from their mainland relatives. The environment will also differ on different islands. But oceanic islands do have certain features in common. Both the vegetation and the fauna will tend to be sparse. Mammals, amphibia and strictly freshwater fish will be the rarest of all the inhabitants. And of the mammals, bats, rats, pigs and small arboreal animals are more likely to be present than others. This precludes the majority of carnivores.

Because of the lack of carnivores on oceanic islands, certain general trends

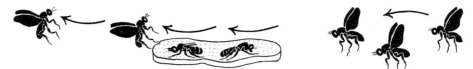

in the evolution of the other island inhabitants can be observed. Birds may gradually evolve into ground-living forms and become large and flightless, like the dodo which lived on the oceanic island of Mauritius where there were no native mammals. Other animals may also get very large, like the giant lizards of the Galapagos Islands and the giant tortoises which are found on several oceanic islands, including the Galapagos and the Seychelles.

Other general tendencies in the fauna of oceanic islands are the evolution of wingless insects, because those that had wings were constantly being blown off the island and were therefore progressively selected out of the population. Birds sometimes lose the bright colours of their mainland relatives, evolving into white or dark forms. And on islands that are very small, the larger inhabitants are liable to extinction because they cannot maintain an adequate population. This reduces the fauna still further.

It is generally agreed that islands can be classified into these two types, oceanic and continental, as Darwin had suggested in 1859. Within this broad classification, islands differ according to whether they are ancient or comparatively new, and their age affects the degree of endemic differentiation of their fauna. This means that a very ancient continental island has many of the characteristics of an oceanic island, because in fact it will have been an isolated island for a very long period.

12

Oceanic Islands

St Helena

St Helena is a volcanic oceanic island some ten miles long and eight wide, lying in deep water in the South Atlantic over 1,000 miles from Africa and considerably more from South America. The ocean currents and the prevailing winds reach the island from the south-west coast of Africa; a branch of the cold Benguela current sweeps up past the island, and the south-east trade winds blow over it.

The most outstanding feature of the fauna of St Helena is the extreme poverty of it. There are no native mammals, reptiles, amphibians nor freshwater fish and the only native land bird is a plover closely related to an African species. Besides the plover, the chief native inhabitants are insects and land molluscs.

There is no climatic reason for the poverty of the fauna as the successful introduction of mammals and insects by man has shown. In fact these introductions may be one of the causes of it. The Portuguese brought goats to the

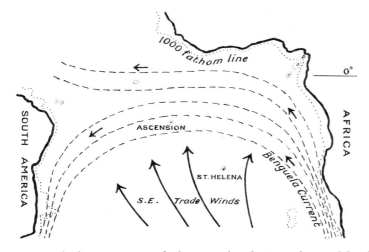

island in 1513 and the ravages of these animals together with the later stripping of the bark from the ebony and redwood trees for the tanning industry, robbed the island of much of its vegetation. As the vegetation was destroyed many animals that depended on it for food and shelter must have died out. This may have reduced the original native insect population in

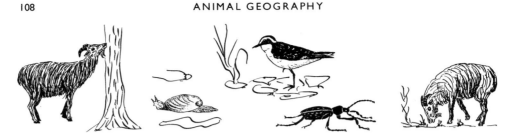

particular and it is possible that others also suffered from the disappearance of the trees.

Even without this recent destruction the St Helena fauna was probably always sparse. Indeed its sparseness and its position in the middle of an ocean indicate that the whole flora and fauna has been acquired by chance immigration from across the sea. The direction of the currents and the prevailing winds suggest Africa as the main source.

The only vertebrate member of the population is the scarcely differentiated African plover. Some weevils are also related to African forms and so are many of the plants. That Africa has supplied some of the colonists is supported by the occasional arrival of driftwood from that continent. Weevils, being wood-boring insects, could have crossed the sea on many occasions buried in floating logs. This method of transport might also account for the St Helena molluscs. Some of the molluscs, however, are so specialized that it is impossible to know where their nearest relatives are to be found, and some weevils have European affinities. Whether molluscs and weevils came in wood by way of Africa and the Benguela current or whether they floated from a more northerly site is impossible to decide, but either of these explanations provides a reasonably acceptable hypothesis. More difficult to explain because the winds and currents cannot so obviously be utilized, is the presence on St Helena of plants that resemble those of South America.

Some biologists have therefore suggested that St Helena was once part of the South Atlantic continent and that it was then colonized from both Africa and South America. This might account for the South American element of the St Helena flora but the fauna has no such distinctly American characteristics, and it seems strange that if St Helena were once part of a lost continent it should have no land vertebrates except a bird whose arrival on the island is comparatively recent.

All things considered, the nature and the composition of the St Helena fauna support the view that the island is oceanic. The bird and the seeds could have come through the air, the invertebrates could have come by sea from Africa. The specialization of the fauna into native genera and species indicates a long residence on the island, amounting to many millions of years.

Galapagos Islands

The Galapagos Islands lie in the Pacific, about 600 miles from the coast of Ecuador. They form a group of fifteen volcanic islands lying across the equator. The coastal areas of the islands are dry and bare, with low thorny bushes and prickly pears. Inland there are sometimes humid forests whose tall trees are covered with ferns and orchids, and sometimes open bare country. These islands have a sparse fauna, but even so it is richer than that of St Helena. It seems to have been easier for animals to cross the 600 miles of Pacific from America to the Galapagos than for them to cross the 1,000 miles of south Atlantic from Africa to St Helena.

Unlike St Helena, the Galapagos Islands have mammals, many land birds, and reptiles, but like St Helena they have no amphibians and no freshwater fish.

The two native mammal species are a bat and the cricetid rice rats. There are about twenty-six different land birds some of which are hardly different from Neotropical species. There is a penguin species, the only one to live in the tropics, and there are unique flightless cormorants. There are also those famous birds, Darwin's finches.

The finches are famous because Darwin studied them when he visited the islands, and as a result of his studies started to think about the reasons for the origin of new species, and eventually formulated his theory of evolution by natural selection. Darwin observed that there were finches of the same two genera spread over all the islands of the Galapagos group, but each island form was recognizably different from all others. In other words it was more or less true to say that each island had its own species. Thus one of the ground finches inhabiting some of the smaller outlying islands shows distinct differences in the proportions of the beak and in its feeding habits. On one island its beak is comparatively slim and it feeds on cactus, on another the beak is altogether broader and heavier and the finch feeds on the ground.

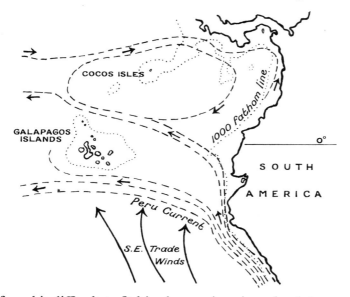

Darwin found it difficult to fit his observations into the then general belief in special creation, and came to the conclusion that the slight differences in the island forms had come about by natural means; that species could evolve into other species. He suggested that when finches colonized a new island, they would be in a slightly different environment from the one they had left, different perhaps in size of territory, in vegetation or in the composition of the other animal inhabitants. Gradually in succeeding generations the finches most fitted to the new surroundings, cactus-eating for instance, would survive and leave offspring whilst the others would die off. Separated from the parent population, the colony would finally evolve into a new species fitted to the new island and different from the original. A further interesting feature of the Galapagos finches is their evolution into forms with many different habits. Owing to the scarcity of other birds on the islands, the

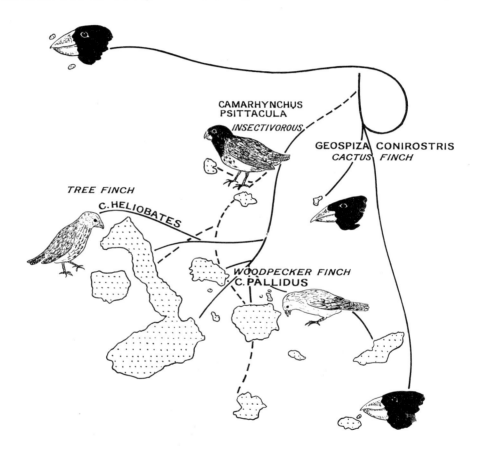

finches have evolved by natural selection from a seed-eating ancestor, not only into cactus-feeders, but also into woodpecker-finches, insectivorous tree finches, mangrove-finches and vegetarian tree finches, differing from one another in beak shape and size, and to a lesser degree in male plumage colour. None of the finches is brightly coloured, having island characteristics of dull plumage, varying in this case from black to light brown.

The giant tortoises, after which the islands are named, also vary from island to island. Unfortunately these lumbering reptiles are on the verge of

extinction. They provided excellent food and oil for the early buccaneers and whalers and for modern settlers. The same fate is overtaking the land iguanas, three foot long lizards that live in burrows and feed almost entirely on cactus. The marine iguanas, however, are still abundant. They live mainly on the shore, sunbathing and feeding on seaweed, while scarlet land crabs run over them eating their parasitic ticks. The other reptiles of the Galapagos, one species of snake, and two of lizards are less remarkable. The snakes and small iguanid lizards are unique species of tropical American genera while the geckos belong to a widely distributed genus.

Several attempts have been made to account for the fauna of the Galapagos by means of a land bridge from South America. True, America has provided most of the fauna, but the very rich South or Central American fauna is only scantily represented on the islands so that it seems more likely that chance sea-crossings can account for the fauna much better than a land bridge. There is no shallow sea area to provide good geological evidence of a land bridge, and the bridge builders are never united in deciding whether to bring the bridge from South America or from Central America.

The vertebrate fauna contains just those animals most likely to have crossed the sea successfully. There are no amphibians, no freshwater fish. Rats and bats are the most successful of mammalian voyagers. The penguin species almost certainly came from the south with the northward-flowing cold Humboldt (or Peruvian) current, but the cormorants probably flew to the islands, only evolving into a flightless species after their arrival. The reptiles, or their eggs, may have come on driftwood, but it also possible that the tortoises floated there in the sea.

There seems no reason to doubt Darwin's classification of the Galapagos Islands as oceanic.

Krakatau

The tiny island group of Krakatau has very different characteristics from St Helena and the Galapagos Islands, for although the islands are volcanic, and genuine oceanic islands, they lie close to land and their fauna is of recent origin.

The three islands lie in the Sunda Straits between Sumatra and Java,

twenty-five miles away from both these large continental islands. Early accounts of the Krakatau Islands describe one large island with three volcanoes on it and two small islands nearby. In 1883 the large island blew up, leaving only a small part of the original above water, and the islands completely covered in hot volcanic ash. Since then the islands have had an uneasy history, sinking and rising and changing their profiles. In spite of this always some part of each of them has remained above the sea during the years since 1883.

When the volcanoes erupted all the organic life of the islands was destroyed. The modern population has, therefore, arrived since then and has arrived from at least twenty-five miles away.

It has been possible to observe the events in the repopulation of these oceanic islands over a period of about forty years.

The zoology of the islands was not known before the eruption, and indeed not until twenty-five years afterwards but since then the large island has been studied on three separate occasions. In 1908 the first brief investigation was made. The supposedly first vegetation, a covering of algal growth over the ground, followed by ferns, had already given way to the third stage in the floral population of volcanic islands, grass wilderness. Animals can only establish themselves in suit-

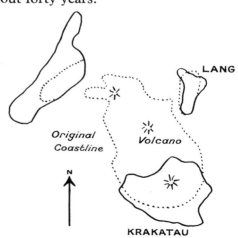

able conditions of vegetation, so that the sequence of their establishment on Krakatau depended on the conditions on the islands as well as the supply of colonists from the nearby islands of Sumatra and Java. Herbivorous animals must establish themselves on an island before carnivores can become part of the population.

By 1908, that is twenty-five years from the beginning of recolonization, there were already thirteen species of birds, including pigeons, kingfishers, orioles and a bulbul, on the large island. There was a species of monitor lizard and a house-gecko as well as 192 species of insects, in which ants and flies predominated. There were thirty-one species of beetle, six of butterflies and two of dragonflies. There were spiders and scorpions, a few woodlice, two species of snail, and very little else.

1908-1921

In 1921 when the next survey was made, there were three times as many species of animals as there had been thirteen years earlier. The grass wilderness was giving way to mixed forest and the first mammals had arrived. These were two species of bats and a species of house rat which had probably come over with a human visitor to the island. The birds had increased to twenty-nine species and included swifts and sunbirds. Another gecko, a skink and a python species had arrived, and the beetles, butterflies and dragonflies had increased enormously in number of species. Also two species of earthworm were now abundant.

Another twelve years later, in 1933, the mixed forest had become dominant, and with the disappearance of much of the grassland, the grass fauna, the gnats and flies for instance, had decreased both in total number and in number of species. But two more bat species had arrived, one an insect-eater. Another rat species, more birds, including flycatchers, another skink and a gecko, and a crocodile species were now on the island. Also almost all the other inhabitants had been joined by new species. Remarkably few forest forms, however, had come to live in the mixed forests of the island although there was now a total of 1,100 species. Only a few insects and a tree-snail represented the tropical forest fauna of the surrounding islands. No mammals, reptiles or leeches had migrated from the neighbouring forests.

Investigation showed that the majority of the colonists had come south from Sumatra, but a few were Javanese. The insects seemed to have been blown to the island, for they were found in the upper air currents. Probably a continuous wave of insects and spiders was being blown across the island and it is not surprising that the number of species increased rapidly once suitable vegetational conditions prevailed. The reptiles may have swum the twenty-five miles, but the earthworms must have come in driftwood or with the birds.

Krakatau is typical of oceanic islands in having no amphibians and no mammals other than bats and rats, in spite of its closeness to land. It has no endemic species for its fauna has not been on the island long enough to have differentiated from that of the nearby islands. It may not seem an ideal oceanic island, but it has been under continuing observation since soon after repopulation began and therefore provides one of the few experiments in oceanic island colonization.

13

Continental Islands

British Isles

The geological history of the British Isles is complex. During the Paleozoic and Mesozoic the northern and western land may have been an outlying part of the large North America-Greenland continent, sharing its fauna. The more southerly parts were frequently below the sea or formed an archipelago of

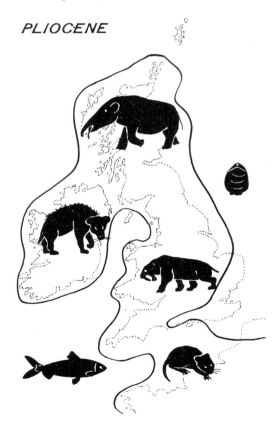

PLIOCENE

islands between the Nearctic and Palearctic continents of the northern hemisphere. Only at the end of the miocene did the British Isles become part of Europe with no land connexions across the North Atlantic (for details see Wills 1951).

The miocene-pliocene fauna was typical of the cooler zones of the northern hemisphere with sabre-tooth tigers, tapirs and mastodons, and Palearctic bears and hyenas. Crocodiles and turtles may have been found still in the rivers of the country and lizards and amphibia can be presumed to have been abundant.

None of this fauna exists in the British Isles today. There are no relics of its early connexion to America with the dubious exception of a freshwater sponge, and there are no pliocene relics of its European fauna. The whole of the British pliocene fauna was wiped out during the pleistocene and the modern fauna has been gained since then. But the pleistocene extinctions were not a simple event and the country was inhabited by several waves of European animals during that epoch.

At the beginning of the pleistocene the climate of the northern hemisphere had become cold, and ice covered the lands in the far north. This cold affected the fauna of the British Isles. Warmth-loving animals moved out or died out and the fauna became typical of a cold temperate region, with otters, red deer, rodents and mastodons. But the cold did not last and as warmer conditions returned so did the sabre-tooths, lions and Old World monkeys. The tapirs, crocodiles and turtles had vanished for ever from the fauna.

A few thousand years later ice covered the British Isles as far south as the Thames and the Severn estuary. This, the second pleistocene glaciation, was the most extensive and was responsible for the elimination of the whole of the earlier British fauna. When the ice retreated again, first reindeer and musk ox, and then monkeys, voles, hyenas, horses, rhinoceroses, pigs, hippos, deer and early Stone Age man, travelled across Doggerland, from Denmark and Germany, and the Channel land, from France, to the British Isles.

This fauna, too, had to give up its tenure of the British Isles when the third ice age made conditions difficult for life. The third ice age was not as rigorous as the second and, although many animals left the country others like the reindeer, musk ox and woolly rhinoceros were able to live on. Again the climate improved and wolves, bears, lions, mammoths, elephants, horses, rhinoceroses, hippos, deer, bison and man came back from Europe. There were also many small rodents in this last interglacial period hardly distinguishable from modern species.

The fourth ice age followed, and forced the fauna out of the British Isles yet again, and many of the pleistocene animals never returned.

Between the retreat of the last ice and the cutting of the Straits of Dover the modern fauna of the British Isles has come in over the Channel land and Doggerland. The last signs of glaciation in the British Isles disappeared only some 9,000 years ago, although the southern part of the land had been habitable before this. A little later the continental land links sank beneath the sea, the final severance from Europe coming when the Straits of Dover were completed between 7,000 and 8,000 years ago. There were therefore only two or three thousand years available for the post-glacial colonization of the British Isles.

Great Britain

A few thousand years to acquire a new fauna is a very short time in terms of animal evolution; seven thousand years of isolation is short in terms of speciation. It is not surprising therefore that the list of modern land vertebrates should be short, the list of endemic species even shorter.

Birds, and in particular sea birds which make up a large proportion of the avifauna, are an exception.

There are shrews, moles, hedgehogs; rabbits and hares; squirrels, cricetids (voles), murid rats and mice, dormice; mustelids (badgers, otters, martens, stoats and weasels), cats; a few cervids (roe deer and red deer), and a few bats. Many European mammals for which the climate seems suitable

are absent; there are no hamsters, lemmings, bears, ibexes, chamois or reindeer; and wolves, beavers and wild boar which were once typical of the land are no longer found.

Reptiles and amphibia are even more poorly represented. There are only three snakes, grass snake, adder and smooth snake and the last is restricted to a few southern counties. There are two species of lacertid lizards, the common brown lizard which is viviparous and the greener sand lizard which lays eggs and is restricted like the smooth snake to the south of the country. A legless lizard, the slow worm *Anguis* completes the reptile fauna.

There are only six species of amphibia, common toad, frog and natterjack amongst the anura, and three species of newt. The edible frog and the marsh frog are recent introductions and there are no midwife toads, no fire toads and no yellow-bellied toads which are common on the other side of the Channel.

There are perch, pike, loach and carp amongst the primary freshwater fish. The char and whitefish which have differentiated into numerous species in British lakes are capable of living in salt water and their problems of colonization are not, therefore, exactly the same as those of the more land-locked vertebrates.

Compared with Europe the modern fauna is sparse, rather less than half the western European species occurring and it has none of the large mammals and reptiles of the pliocene. There are only six species of amphibians compared with twelve in France and the Low Countries; only about 50 species of land mammal compared with nearly 100 in Germany and 70 in Scandinavia. The short time available for the colonization of the country must be the main cause of the omissions, but the climate may also be responsible for the absence of some European species such as the green lizard and European tree frog for example.

In seven thousand years the portion of the European fauna isolated in Great Britain has shown scarcely any evolutionary change. There is only one endemic species of land vertebrate, the red grouse, *Lagopus scoticus*, and the specific status of this is sometimes disputed. But although they may not have attained specific rank, several British land animals differ recognizably from their European relatives. The British red squirrel can be distinguished from the European red squirrel by the lighter colour of its tail. The British coal-tit has greenish upper parts, the European coal-tit, slate-blue. There has been some differentiation in seven thousand years and this is indicated in fauna lists by the addition of a sub-specific or varietal name to the usual binomial.

As far as is known the modern fauna of Great Britain was acquired mainly across land from its continental parent. It is typical of a recent island in the lack of differentiation of its fauna.

Whether any members of the fauna are even more recent immigrants from

across the sea is difficult to tell when the fauna is so little differentiated. Some birds are certainly recent, some having arrived and established themselves in historic times. Butterflies, too, are known to migrate to England; the Camberwell Beauty from Scandinavia; clouded yellows and several blues from France. There is no substantiated case of any other animals having crossed the sea and established breeding colonies in the country, but the possibility cannot be ruled out.

Ireland

Ireland is a second-hand continental island. It acquired its fauna by way of Great Britain and became an independent island only recently, after it was isolated from the continent. Therefore the fauna is scanty, only about half the British species being represented, and the differentiation of the sparse fauna into Irish species has been slight. The Irish hare is sometimes differentiated from the British and European hares at the species level but most of the differences between Irish and British populations are so small as to merit subspecific rank at the most. Just as the British coal-tit differs from the European, so the Irish population can be distinguished from the British by slight colour differences of the cheeks. There is a subspecies of the red grouse and several invertebrates have distinctive characteristics.

Only eight families of mammals are represented in Ireland and with the exception of a deer they are all small and insignificant. There are shrews, hedgehogs, rabbits, squirrels, murids, foxes, mustelids and cervids. Only one of the four British shrews, the pygmy shrew, and one of the two British deer, the red deer, are found in Ireland. There are no moles, no cricetids, no dormice and no wild cats although all but the last of these families is abundant in Great Britain.

Of the twelve species of British reptiles and amphibians, itself a meagre number, only three are found in Ireland. These are the natterjack toad which is confined to a small area of Kerry, the common newt and the viviparous lizard. The frog which is a flourishing member of the fauna today was introduced in 1696. There are no snakes in Ireland but it is doubtful whether it was St Patrick who was responsible for banishing them.

Of the sixteen species of strictly freshwater British fish, only eight have reached Ireland.

Like the rest of the British Isles, Ireland was covered by ice during the second and third ice ages and it is unlikely that any of the pliocene fauna could have survived the severe climate. The animals of Ireland are therefore post-glacial arrivals. But unlike those of Great Britain, the Irish migrants had no direct land connexion from Europe to Ireland after the ice had left. As already indicated, their journey to Ireland passed through the greater part of Britain.

In the years immediately after the retreat of the ice, Ireland was connected to Scotland by a neck of land stretching from north-east Donegal to the islands of Islay and Jura in south-west Scotland. This was probably the only connexion that Ireland had and it lasted longer than the Channel land connecting Great Britain to the rest of Europe.

Ireland was therefore open to colonization by land for several thousand years and yet only about half the British animals reached Ireland. There was probably more than one reason for the sieving effect, but one of the most important must have been the situation of the land connexion. Being in the north of Great Britain, animals had to spread through almost the whole of the large island before they could gain access to Ireland. Some animals were too late arriving in Great Britain to travel as far north before the corridor was lost, others, even early arrivals, never spread far north for climatic or vegetational reasons and were therefore never eligible for the Irish journey.

The cricetid bank vole *Clethrionomys glareolus* and the dormouse were probably too late to reach Ireland. The dormouse is still confined to England and Wales but the bank vole is widespread over the whole of Great Britain, ranging as far west as some of the islands off the coast of Scotland.

The smooth snake and the sand lizard are confined, probably for climatic reasons, to the southern part of England where they are only found on dry

heaths and open woodland. The egg-laying habit of the sand lizard may be an important factor in its limitation, for the viviparous common lizard has spread over the whole of the British Isles. Whatever the precise reason, the range of the smooth snake and sand lizard has never extended northwards far enough to cover the Islay-Donegal land bridge.

It is more difficult to understand the absence of the mole, the adder, the toad and the newts. They are widespread over Great Britain today reaching to the north and west coasts of Scotland. It is possible, but seems unlikely, that all these vertebrates were too late arriving in Great Britain to extend their range northwards in time. However some of them may have been slow in spreading and others may have been prevented from crossing to Ireland by climatic factors. The adder may have extended its range far northwards only recently and the wide-ranging mole may have been stopped by the physical conditions of the land connexion itself. Moles usually burrow in well-drained soil and the soil of the land may have been too water-logged, frozen, or even too stoney.

There is yet another way of accounting for some of the gaps in the Irish fauna. Ireland is a comparatively small island, it has a uniform climate and little differentiation of vegetation. Some of the missing British vertebrates may have succeeded in colonizing Ireland but were unable to maintain them-

selves either because of the climate or because the populations were too small. It is possible that the climate was unsuitable for moles and perhaps also for grass snakes, so that they never established breeding colonies on the Irish side of the land corridor. The newts and toads on the other hand may have been successful initially but died out because they were unable to maintain big enough colonies. This last suggestion is supported to some extent by the present-day distribution of the natterjack in Ireland. Its small range in the far south-west is indicative of only small success, or approaching extinction.

The poverty of the Irish fauna can therefore be accounted for by its northerly connexion with another poorly populated island, its climate and its small area. Ireland has a selected portion of the British fauna. But there are four Irish animals which do not occur in Great Britain and although they are none of them vertebrates their unusual type of distribution is worth noticing. An earthworm, a woodlouse, a spotted slug and a moth have a small southerly range in Ireland and are otherwise only found in south-western Europe.

To explain this so-called Lusitanian element in the Irish fauna, Edward Forbes in 1846 suggested a land bridge from Portugal to Ireland. Forbes was one of the first scientific biogeographers but he was also a keen bridge builder. A pleistocene Lusitanian bridge is not acceptable on geological grounds and it seems likely that the Lusitanian species were once more widespread than they are today. They may even be survivals from the last interglacial period, becoming restricted in range as the climate of the fourth ice age deteriorated. The southerly part of Ireland was not ice-covered during the fourth, and last, glaciation and might have provided a refuge for these few survivors from a more temperate climate.

14

Ancient Islands

New Zealand

New Zealand lies about 1,000 miles east-south-east of Australia, in the south temperate zone. Within the 100,000 square miles of its two main islands there

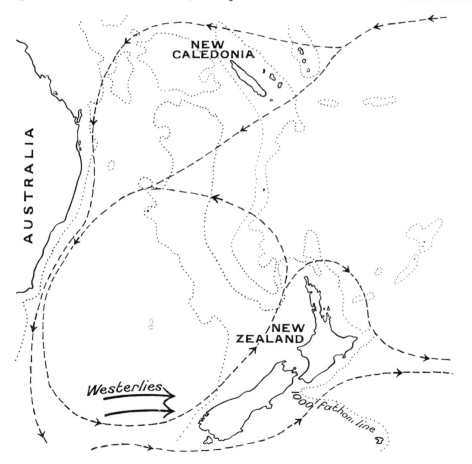

are both mountains and open country and a wide variety of temperate vegetation. Geologically it is complex, with extinct volcanoes as well as sedimentary rocks.

The fauna of New Zealand, like that of other islands, is characterized both by the absence of many animals and by the differentiation of those that are there. But typical of New Zealand itself are relics of old vertebrate stocks and a high proportion of flightless birds, half of which are extinct.

There are no native mammals except two families of bats, one of which is endemic and one Australian. There are no climatic reasons for this deficiency of mammals because those recently introduced have been a great success. But correlated with the absence of mammals, and partially compensating for it in the fauna, is the unique array of flightless birds. There are several species of small kiwis, birds with soft, hair-like feathers and much reduced wings. Their sense of smell is highly developed which may be a primitive feature or connected with their nocturnal habits and diet of worms and insects. There are flightless rails and the kakapo, or owl parrot. The kakapo is hunted for food and although it lives in secluded places and only comes out at night, its numbers are becoming dangerously low. All the other flightless birds of New Zealand have already become extinct. There were twenty species of giant moas, some as large as ostriches. They lived in New Zealand from the end of the miocene until the coming of the Polynesians. There was also a flightless goose and a flightless wren, the only flightless perching bird known.

Of the flying birds, some belong to world-wide families, a few are Australian and there are two small endemic families; New Zealand wrens and the wattlebirds. A remarkable New Zealand bird is the kea, a genus of large parrot, whose habits have changed in an extraordinary way since the introduction into New Zealand of mammals, in this case, sheep. The kea sits on the sheep's back, tearing holes in its flesh to expose the kidney fat which it eats. Before there were any sheep the kea fed on insects, scratched out of crannies with its powerful beak.

New Zealand has geckos and skinks but no turtles or snakes except marine ones. But *Sphenodon*, the last surviving member of an order that disappeared from the fossil record a hundred million years ago, still lives in New Zealand. *Sphenodon*, the tuatara, looks like a large lizard with an overhanging beaked jaw. It is more or less amphibious and its eggs, laid in the sand, take over a year to hatch.

There is only one amphibian proper, the curious frog, *Liopelma*, with tail muscles, related today only to *Ascaphus* of North America.

There are no strictly freshwater fish.

Clearly, the vertebrate fauna is an ancient one containing as it does a relict reptile unknown elsewhere since the cretaceous, and a relict frog surviving only locally today. The fauna is also highly endemic, but at all levels, from an

endemic order (*Sphenodon*), through families (birds and bats), to many endemic genera (amphibian, reptiles and birds) and species (birds and bats).

Where did this fauna originate and how did it get to New Zealand? How long has New Zealand been an island?

New Zealand could have had a connexion with Australia in Mesozoic days. A narrow strip of shallow sea runs from the north-west corner of New Zealand to the north-east corner of Australia, a distance of some 1,000 miles, and this may indicate the site of a Mesozoic or Paleozoic land bridge. Alternatively, according to the drift theory, New Zealand may have been nearer to Australia and in a more southerly latitude, near the Antarctic, in those days.

If any of these land connexions existed, the archaic animals of New Zealand, which were once widespread through the northern continents in the Mesozoic, might have migrated to New Zealand at that time, across land. Invertebrates, *Liopelma* and *Sphenodon* might have used such a route. But any connexion would have been terminated by the jurassic at the latest, preventing the influx of other amphibia and reptiles, birds and mammals, which do not occur on New Zealand.

Thus although New Zealand may be an ancient continental island, the later part of its vertebrate fauna has been acquired like that of oceanic islands from across the sea.

There is no particular reason to suppose the flightless birds needed a land bridge, for there is no evidence that they were already flightless on arrival. No fossil flightless relatives are known from Australia, and it is therefore more likely that they arrived as flying forms at the beginning of the Cenozoic. In the absence of mammals and other predators they then evolved into the many ground-living forms.

The geckos and skinks are only generically different from Asiatic forms and are therefore obviously more recent casual arrivals from across the sea.

New Zealand then may be an ancient continental island with a relict fauna characteristic of an early phase of land connexion, but with a later fauna characteristic of an isolated island. Alternatively, because its relict fauna is so small, it may have been an island for the whole of the Mesozoic and Cenozoic, receiving all its vertebrate fauna, its tuatara and its *Liopelma*, its birds and its lizards at varying times from across the sea.

An island since the end of the Mesozoic, the earlier status of New Zealand is undecided.

Madagascar

The problem is similar to that of New Zealand, but whereas most zoo-geographers believe that New Zealand has been independent for at least the whole of the Cenozoic, there are many who believe that Madagascar was part of Africa until paleocene or even eocene days, for Madagascar has native mammals.

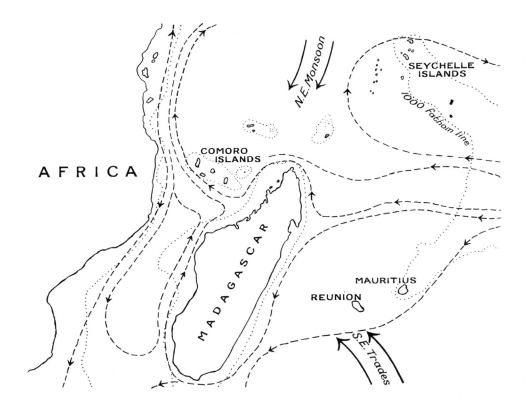

Madagascar is a large island, larger in area than the islands that make up New Zealand, and lies 260 miles from the east coast of Africa, in the Indian Ocean.

Geologically it is an old island, a granite plateau runs down the centre with scattered volcanic peaks. There are dense fern forests and bamboo thickets, tropical swamps, bare rock country and sand dunes. Many of the native mammals are forest-dwellers and it is probable that Madagascar was once more thickly forested than it is today.

Of modern Malagasy mammals, shrews, murid mice and an African bush pig are almost certainly recent human introductions. There are many bats, with both Ethiopian and Oriental affinities. The remaining seventy-five species of land mammal represent only four orders: Insectivora, Primates, Rodentia and Carnivora.

There is one family of insectivores, Tenrecidae, with thirty species. The family is unknown elsewhere, but in Madagascar it has paralleled more widely distributed insectivore forms. There are large spiny hedgehog tenrecs up to eighteen inches long, long-tailed mouse tenrecs, rice tenrecs with stout digging claws like moles, and water tenrecs with webbed feet and flattened tails.

The primates are represented by three families of lemurs. Two families contain mainly monkey-like animals, the well-known ring-tail lemurs, sportive lemurs, gentle lemurs, woolly lemurs and silky lemurs. The third is the aye-aye with chisel teeth specialized for a rodent life in the dense bamboo forests.

A subfamily of cricetid rats and several endemic genera of civets complete the modern mammal fauna. In addition a pygmy hippopotamus is known from pleistocene deposits.

There are plenty of birds on Madagascar. A large number are endemic, the majority of the rest African and a few are Oriental. Four families are confined to Madagascar, two of them flightless, the giant elephant birds which are extinct, and the rail-like mesoenatids. The other two families are philepittas and the black, white and chestnut vangas. Both forest families, the philepittas are fruit-eaters and the vangas feed mainly on chameleons and insects. But in contrast to diversity and mixed origin, there is a conspicuous lack in the avifauna of certain families. There are no ostriches, secretary-birds, touracos or mousebirds all of which are typical of Africa. There are no hornbills or woodpeckers, which are widely distributed through the Ethiopian and Oriental regions.

The reptile and amphibian component of the fauna follows the same pattern: absence of several African and Oriental families, a number of endemic genera, and the majority of the rest African. Thus agamid lizards (widespread through the Old World) and all poisonous snakes, pipid toads and common toads are absent. There are instead, pelomedusid side-neck turtles, land turtles, geckos, chameleons, skinks and spiny lizards which are mainly African and Old World families. But there are also land iguanids, a typically American family.

The polypedatid tree frogs, typical of Old World tropics, are well represented on Madagascar. Two genera are shared with Africa, one, *Polypedates*, with the Oriental region, and four are endemic.

There are no strictly freshwater fish in Madagascar.

The vertebrate fauna of Madagascar is a good deal more varied than that of New Zealand, but nevertheless it lacks the variety of its nearest neighbour, Africa. There are no large mammals or freshwater fish, and many families of Old World birds, reptiles and amphibia are lacking. In the main the fauna shows African affinities, but undoubtedly some of the birds and amphibia are more closely related to Oriental forms.

The many endemic families and genera on Madagascar indicate a long period of isolation, but as many of the endemic families are mammals and might need a land connexion, the isolation may date only from the paleocene or eocene. Madagascar may be, in fact, an ancient continental island and its continental connexions might be either with Africa or India, or both.

The continent of Lemuria, already discussed, would provide Madagascar with continental connexions in both an easterly and a westerly direction. Across Lemuria could have travelled lemurs, civets, some birds, chameleons and polypedatid frogs, to be shared by all. But Lemuria does not provide a satisfactory explanation of the animals it was particularly invented for, the lemurs, and has therefore already been dismissed.

It is more generally believed that Madagascar had a land connexion only with Africa during the Mesozoic. This would permit the African amphibia, reptiles, birds and earliest mammals to enter Madagascar by land. Thus some ancestral insectivores, lemurs, viverrids, cricetids, side-neck turtles and chameleons would have travelled to Madagascar between the cretaceous and the eocene. The difficulty with this theory is that there is no evidence that lemurs, viverrids and cricetids were in Africa all together or early enough. To put all the mammalian orders on the island by land, means waiting at least until eocene or even later days.

If this late date is correct, it is remarkable that, as far as is known, no condylarths populated Madagascar, nor any of the freshwater fish nor pipid toads which must have been in Africa then.

If land mammals require a land bridge to Madagascar it might be thought that amphibians needed one even more. But the polypedatid tree frogs are thought to have been late anuran derivatives, oligocene or miocene, and would need a late land bridge, presumably to both Africa and the Oriental region.

Bridges are rearing up in all directions at all times, and the geological evidence is against them.

The Mozambique channel between Africa and Madagascar is deep and the rock formations on either side suggest that the channel dates from at least triassic days. It may not always have been as wide as it is today but it seems to have been continuous. If this is confirmed, Madagascar must be considered to have been an island since the early Mesozoic.

The status of ancient island is supported by the composition of the Malagasy fauna. Many of the families are old, but neither are they all of the same age nor are they fully representative of an old African fauna. Some mammals and many amphibians are missing. Those that are found in Madagascar are mainly small and arboreal, just those that are most likely to sweep across the sea on 'rafts'. A short sea-crossing, selective by its very nature, can account for the endemic and African elements in the fauna, for

their apparent haphazard selection and for their varying ages. Tenrecs and the ancestors of the flightless elephant birds may have been early arrivals, followed by lemurs and cricetids, with civets, polypedatids and the pygmy hippo much later arrivals.

There is still the Oriental fraction of the fauna to be accounted for, and the American iguanids and the side-neck turtles.

The Oriental genera are comparatively few; four genera of birds, a few reptiles and frogs. It seems reasonable to suppose that the islands of the Indian Ocean that lie north of Madagascar may have been more numerous or more extensive at earlier dates and may have supplied stepping stones for these occasional migrants.

The iguanids and the side-neck turtles are a more difficult problem. Two genera of Malagasy side-necks are African but the third, *Podocnemis*, is South American. It is not easy to understand why American iguanids and a South American turtle should inhabit Madagascar. But fossil evidence suggests that the South American turtle was recently an inhabitant of Africa as well, and its presence in Madagascar is therefore no more surprising than that of any other African turtle. There is no such easy explanation of the iguanids. They are thought to have been a widespread family in the Mesozoic and the Malagasy genera may be relicts of an early African reptile fauna.

It is no easier to decide the status of Madagascar than it is to make a decision over New Zealand.

Madagascar has more mammals, reptiles and amphibia than New Zealand but they are not less highly differentiated, and, like New Zealand, is without primary freshwater fish. This might mean that Madagascar had a land connexion with Africa until the beginning of the Cenozoic to acquire its mammals and amphibia after which it became a discrete island, or it may mean, and the lack of freshwater fish and geological evidence supports this, that Madagascar has been an island since the early Mesozoic. If it has been an island for this length of time it should resemble New Zealand. In the specialization of its fauna it does resemble New Zealand but, in marked contrast to New Zealand, it has a flourishing fauna of mammals and amphibia. However, as the native mammals represent only four orders and the amphians only four families, it is likely that they derived from very few initial immigrants. One tenrec and one lemur colonization would have been enough to found the modern populations. The distance from Africa to Madagascar is only a quarter of that between Australia and New Zealand and may not have been as effective a sea barrier against amphibians and mammals.

Like New Zealand then, Madagascar has probably been isolated since the Mesozoic receiving its modern fauna from across the sea, but what happened before that cannot yet be decided.

Conclusion

Animal populations evolve, spread and contract, limited by the geography, climate, flora, fauna and their own inherent propensities, and this complex and changing interaction of physical and living factors operates at all times. The study of present-day animal geography, the distribution of animals in space, is a cross-section taken at an arbitrary moment in time, of a process which has been going on steadily through the past. Only by understanding the changing scene, by combining and understanding the changes in time, the evolution, with changes in space, the distribution, can the apparent anomalies of animal geography be understood. The causes of distribution in the present are complex; they have been equally complex in the past.

Book List

ALLEE, W. C., EMERSON, A. E., PARK, O., PARK, T. & SCHMIDT, K. P.: *Principles of Animal Ecology*: Philadelphia & London, 1949

BARTHOLOMEW, J. G., CLARKE, W. E. & GRIMSHAW, P. H.: *Atlas of Zoogeography*: Edinburgh, 1911

BEAUFORT, L. F. DE: *Zoogeography of the Land and Inland Waters*: London, 1951

BEIRNE, B. P.: *The Origin and History of the British Fauna*: London, 1952

COLBERT, E. H.: *Evolution of the Vertebrates*: New York, 1955

DARLINGTON, P. J.: *Zoogeography*: New York & London, 1957

DUNBAR, C. O.: *Historical Geology*: New York, 1949

EKMAN, S.: *Zoogeography of the Sea*: London, 1953

ELTON, C. S.: *The Ecology of Invasions by Animals and Plants*: London, 1958

HESSE, R., ALLEE, W. C. & SCHMIDT, K. P.: *Ecological Animal Geography*, 2nd edit: New York & London, 1951

HOLMES, A.: A revised geological time-scale: *Trans. Edin. Geol. Soc.* 17:183: 1960

IRVING, E. & GREEN, R.: Paleomagnetic evidence from the cretaceous and cainozoic: *Nature* 179:1064: 1957

JOLEAUD, L.: *Atlas de Paléogéographie*: Paris, 1939

LACK, D.: *Darwin's Finches*: Cambridge, 1947

MATTHEWS, L. HARRISON: *British Mammals:* London, 1952

MAYR, E.: Wallace's line in the light of recent zoogeographic studies: *Q.Rev.Biol.* 19:1: 1944

MAYR, E. (editor): The problem of land connections across the south atlantic with special reference to the mesozoic: *Bull.Amer.Mus.Nat.Hist.* 99:85: 1952

MAYR, E. & AMADON, D.: A classification of recent birds: *Amer.Mus. Novitates* 1496: 1951

MOREAU, R. E.: Africa since the mesozoic: *Proc.Zool.Soc.London* 121:869: 1952

NEAVERSON, E.: *Stratigraphical Paleontology*, 2nd edit: London, 1955

NOBLE, G. K.: *The Biology of the Amphibia*: New York, 1931

ROMER, A. S.: *Vertebrate Paleontology*, 2nd edit: Chicago, 1945

ROMER, A. S.: *Osteology of the Reptiles*: Chicago, 1956

SCOTT, W. B.: *A History of the Land Mammals in the Western Hemisphere*, Revised edit: New York, 1937

SIMPSON, G. G.: Antarctica as a faunal migration route: *Proc.6thPacific Sci.Congr.* 2:755: 1940

SIMPSON, G. G.: Mammals and the nature of continents: *Amer.J.Sci.* 241:1: 1943

SIMPSON, G. G.: The principles of classification and a classification of mammals: *Bull.Amer.Mus.Nat.Hist.* 85:1: 1945

SIMPSON, G. G.: Evolution, interchange, and resemblance of the North American and Eurasian cenozoic mammalian faunas: *Evolution* 1:218: 1947

SIMPSON, G. G.: History of the fauna of Latin America: *Amer.Scientist* 38: 361: 1950

SIMPSON, G. G.: *Life of the Past*: New Haven & Oxford, 1953
SIMPSON, G. G.: *Evolution and Geography*: Eugene, Oregon, 1953
TOIT, A. L. DU: *Our Wandering Continents*: Edinburgh, 1937
WALLACE, A. R.: *Geographical Distribution of Animals*: London, 1876
WALLACE, A. R.: *Island Life*: London, 1880
WILLS, L. J.: *Paleogeographical Atlas*: London & Glasgow, 1951
WOOD, A. E.: A revised classification of the rodents: *J. Mammal.* 36:165: 1955
YOUNG, J. Z.: *The Life of Vertebrates*: Oxford, 1950

Index to Regional Illustrations

Read the animals on the illustrations from *left to right, from top to bottom*

PAGE 38. Jumping mouse *Zapus*, Gila monster *Heloderma*, prairie dog *Cynomys*, marmot *Marmota*, rattlesnake *Crotalus*, horned lizard *Phrynosoma*, kangaroo rat *Dipodomys*.

39. Jerboa *Jaculus*, gazelle *Addax*, ground squirrel *Xerus*, horned viper *Cerastes*, gerbil *Acomys*, spiny-tailed lizard *Cordylus*.

67. Side-neck turtle *Podocnemis*, side-neck turtle *Podocnemis*, lungfish *Ceratodus*, insectivore *Puercolestes*, insectivore *Deltatheridium*, condylarth *Oxyacodon*, lungfish *Ceratodus*, marsupial *Eodelphis*, creodont *Oxyclaenus*.

68. Primate *Adapis*, side-neck turtle *Podocnemis*, marsupial *Peratherium*, insectivore *Adapisorex*, creodont *Sinopa*, side-neck turtle *Taphrosphys*, camelid *Protylopus*, rhinocerotid *Prohyracodon*, condylarth *Hyopsodus*, insectivore *Phenacops*, sciuromorph rodent *Plesiarctomys*, sciuromorph rodent *Paramys*, insectivore *Adapisorex*, marsupial *Peratherium*, primate *Notharctus*, creodont *Sinopa*.

70. Rhinocerotid *Aceratherium*, side-neck turtle *Podocnemis*, marsupial *Peratherium*, insectivore *Talpa*, side-neck turtle *Taphrosphys*, pig *Paleochoerus*, beaver *Paleocastor*, camelid *Oxydactylus*, cat *Pseudaelurus*, insectivore *Proscalops*, marsupial *Peratherium*, proboscid *Gomphotherium*, deer *Blastomeryx*, peccary *Perchoerus*, deer *Amphitragulus*, beaver *Steneofiber*, cat *Archaelurus*, rhinocerotid *Diceratherium*.

71. Rhinocerotid *Dicerorhinus*, side-neck turtle *Paralichelys*, ape *Dryopithecus*, elephant *Stegolophodon*, insectivore *Talpa*, beaver *Castor*, pig *Sus*, beaver *Castor*, insectivore *Scalopus*, camelid *Procamelus*, hyena *Crocuta*, deer *Cervus*, deer *Cranioceras*, peccary *Prosthennops*, cat *Pseudaelurus*, murid *Parapodemus*, cat *Felis*, proboscid *Tetralophodon*, pronghorn *Sphenophalos*.

75. Side-neck turtle *Podocnemis*, lungfish *Ceratodus*, condylarth *Oxyacodon*, marsupial *Eodelphis*, lungfish *Ceratodus*, side-neck turtle *Podocnemis*.

76. Side-neck turtle *Taphrosphys*, sciuromorph, primate, condylarth *Didolodus*, armadillo *Utaetus*, notoungulate *Thomashuxleya*, marsupial *Ideodelphys*, side-neck turtle *Podocnemis*.

77. Camelid *Oxydactylus*, procyonid *Alectocyon*, side-neck turtle *Taphrosphys*, notoungulate *Adinotherium*, cebid *Homunculus*, marsupial *Prothylacinus*, chinchilla *Perimys*, litoptern *Thoatherium*, armadillo *Peltephilus*, side-neck turtle *Podocnemis*, marsupial *Microbiotherium*.

78. Peccary *Prosthennops*, camelid *Procamelus*, raccoon *Procyon*, notoungulate *Trigodon*, opossum *Didelphis*, carnivorous marsupial *Thylacosmilus*, litoptern *Promacrauchenia*, armadillo *Chaetophractus*, side-neck turtle *Podocnemis*, chinchilla *Euphilus*.

79. Side-neck turtle *Podocnemis*, lungfish *Ceratodus*, condylarth, lungfish *Ceratodus*, insectivore, side-neck turtle *Podocnemis*.

80. Primate *Adapis*, side-neck turtle *Podocnemis*, creodont *Sinopa*, proboscid *Moeritherium*, side-neck turtle *Podocnemis*, insectivore.

81. Side-neck turtle *Podocnemis*, rhinocerotid *Aceratherium*, *Arsinoitherium*, pig *Paleochoerus*, ape *Proconsul*, side-neck turtle *Podocnemis*, proboscid *Gomphotherium*, cat *Pseudaelurus*, creodont *Hyaenodon*, lungfish *Protopterus*.

82. Ancestral tree shrew, creodont *Propterodon*, side-neck turtle *Podocnemis*, primate *Pondaungia*, rhinocerotid *Fostercooperia*.

83. Proboscid *Gomphotherium*, deer *Amphitragulus*, side-neck turtle, cat *Hyaenaelurus*, pig *Paleochoerus*, rhinocerotid *Aceratherium*.

84. Hyena *Crocuta*, ape *Bramapithecus*, side-neck turtle *Shweboemys*, elephant *Stegodon*, murid *Parapodemus*, cat *Panthera*, rhinocerotid *Rhinoceros*, deer *Cervus*, pig *Sus*.

85. Condylarth *Phenacolophus*, insectivore, marsupial, lungfish *Ceratodus*.

86. Deer *Cervus*, pig *Sus*, bat, agamid lizard, murid, boid snake, phalangerid marsupial *Wynyardia*, lungfish *Neoceratodus*.

104. *Tarsius*, cockatoo *Cacatua*, dwarf buffalo *Anoa*, cuscus *Phalanger*, starling *Basilornis*, *Babirussa*, *Cynopithecus*, *Papilio*.

Index